Monographs on Inorganic and Physical Chemistry

EDITED BY

ALEXANDER FINDLAY, M.A., D.Sc., F.I.C.

Professor of Chemistry, University College of Wales, Aberystwyth.

To those engaged in guiding the reading of advanced students of chemistry, the difficulty of obtaining adequately summarised accounts of the progress made in recent years, more especially along certain of the more actively pursued lines of advance, becomes ever more acutely felt. So great has now become the volume of chemical investigation, and so numerous the channels of its publication, that not only the Honours Student but also the worker desirous of undertaking Research in one or other department of his subject, feels it a growing difficulty to become *au fait* with the present state of the more important and more strenuously cultivated regions of his Science. To assist these two classes of students—those reading for an Honours Degree, and those undertaking Research—is the main aim of the present Series of Monographs.

In this Series of Monographs it is hoped to place before advanced students of Chemistry, accounts of certain sections of Inorganic and Physical Chemistry fuller and more extended in scope than can be obtained in ordinary text-books. Exhaustive treatment of the different subjects, however, so far as concerns work important in its time but now only of historical interest, will not be attempted; the chief attention will be given to recent investigations.

Arrangements have already been made to publish the following monographs, and should these prove themselves to be of value, others will be issued from time to time.

THE CHEMISTRY OF THE RADIO-ELEMENTS.
By FREDERICK SODDY, F.R.S., of the University of Glasgow. 8vo.

    Part I. 2s. 6d. net.
    Part II. The Radio-Elements and the Periodic Law. 2s. net.
    Parts I and II in One Volume. Price 4s. net.

PER-ACIDS AND THEIR SALTS.
By T. SLATER PRICE, D.Sc., of the Municipal Technical School, Birmingham. 8vo. 3s. net.

OSMOTIC PRESSURE.
By ALEXANDER FINDLAY, D.Sc., Editor of this Series. 8vo. 2s. 6d. net.

INTERMETALLIC COMPOUNDS.
By CECIL H. DESCH, D.Sc., of the University of Glasgow. 8vo. 3s.

THE VISCOSITY OF LIQUIDS.
By ALBERT ERNEST DUNSTAN, D.Sc. (Lond.), Head of the Chemical Department, East Ham Technical College, and FERDINAND BERNARD THOLE, B.Sc. (Lond.), Lecturer on Organic Chemistry, East Ham Technical College. 8vo. 3s.

ELECTROLYTIC DISSOCIATION THEORY.
By J. C. PHILIP, D.Sc., of the Chemistry Department, Imperial College of Science and Technology, South Kensington.

THE PHYSICAL CHEMISTRY OF FLAMES.
By J. E. COATES, M.Sc., of the Chemistry Department, The University of Birmingham.

CLAYS.
By J. W. MELLOR, D.Sc.

CATALYSIS OF GAS REACTIONS.
By D. L. CHAPMAN, M.A., Jesus College, Oxford.

THE ELECTRO-CHEMISTRY OF NON-AQUEOUS SOLUTIONS.
By JAMES W. MACBAIN, Ph.D., of the Chemistry Department, The University, Bristol.

CATALYSIS IN LIQUID SYSTEMS.
By GEORGE SENTER, D.Sc., of St. Mary's Hospital Medical School, London.

MOLECULAR ASSOCIATION.
By W. E. S. TURNER, D.Sc., of the Chemistry Department, The University, Sheffield.

HYDR TES IN SOLUTION.
By Professor E. A. WASHBURN, of the University of Illinois, Urbana, Illinois.

LONGMANS, GREEN & CO., 39 PATERNOSTER ROW, LONDON, E.C.

NEW YORK, BOMBAY AND CALCUTTA.

MONOGRAPHS ON INORGANIC AND PHYSICAL CHEMISTRY

EDITED BY ALEXANDER FINDLAY, D.Sc.

INTERMETALLIC COMPOUNDS

INTERMETALLIC
COMPOUNDS

BY

CECIL H. DESCH, D.Sc., Ph.D., F.I.C.

GRAHAM YOUNG LECTURER IN METALLURGICAL CHEMISTRY IN THE UNIVERSITY
OF GLASGOW

WITH SEVENTEEN FIGURES

LONGMANS, GREEN AND CO.

39 PATERNOSTER ROW, LONDON

NEW YORK, BOMBAY AND CALCUTTA

1914

CONTENTS

Specific Volume, 39. Hardness, 42. Electrical Conductivity, 45. Thermal Conductivity, 54. Thermoelectric Power, 55. Electrolytic Potential, 57. Heat of Formation, 61. Specific Heat, 63. Optical Properties, 63. Photo-electric Properties, 64. Magnetic Properties, 65. Hall and Nernst Effects, 74. Crystalline Form, 75.

The Vapour Phase of Metallic Systems, 84.

THE CHEMICAL NATURE OF INTERMETALLIC COMPOUNDS, 88. Compounds of Metals of Group I with the Metals of Other Groups, 94. Compounds of the Metals of Group II with Metals of Other Groups, 98. Compounds of the Metals of Group III with Metals of Other Groups, 99. Compounds of the Metals of Group IV with Metals of Other Groups, 100. Compounds of the Elements of Group V

CONTENTS.

CHAPTER I.

INTRODUCTION.

THE fact that metals are capable of forming chemical combinations among themselves has only gradually received recognition. Alloys were generally regarded as mixtures, and their variability of composition was cited during the controversy between Proust and Berthollet as to the definiteness of proportions in chemical combination. The earliest suggestions that compounds might be present in certain alloys were based on thermal observations. In the course of a series of careful determinations of the freezing-points of some fusible alloys, Rudberg[1] observed that the thermometer generally showed two arrests during cooling, the first depending on the composition of the alloy, whilst the position of the second was constant throughout any one series. Such a constant lower freezing-point was observed in the series lead-tin, lead-bismuth, bismuth-tin, bismuth-zinc, and zinc-tin, and was attributed to the formation of a compound or " chemical alloy ". Formulæ were assigned to several of these supposed compounds, which are now known to be eutectic mixtures. This view long survived, and received much support from the work of Levol,[2] who observed that liquation occurred in all alloys of silver and copper, with the exception of that containing 71·89 per cent of silver, which he therefore assumed to be a definite compound, with the formula Ag_3Cu_2 (using the modern atomic weights). At a much later date, Guthrie, in the course of a study of salt solutions, observed the occurrence of a constant minimum freezing-point in many series, and this he regarded as due to the chemical combination of salt and water to form a " cryohydrate," stable only at low temperatures.[3] This erroneous view unfortunately prevailed, although the true nature of the minimum, as the point of intersection of the ice curve and the salt solubility curve, had been shown earlier by Rüdorff.[4] Guthrie afterwards conducted extensive researches into the freezing of salt solutions and

I

alloys, and, correcting his former view, introduced the word
"eutectic" to denote the mixtures of constant minimum freezing-point.[5]   The assigning of a chemical formula to a eutectic alloy is to be found even in some recent work, where
freezing-point curves have been determined experimentally
without a proper comprehension of their meaning.   This false
assumption of chemical compounds is not to be confused with the
quite legitimate attempt to discover definite atomic or molecular ratios in eutectic mixtures, the heterogeneous nature of the
eutectic equilibrium being fully recognized.[6]

A more promising approach was made in a different way.   So
far back as 1839, Karsten[7] observed the change of colour of
alloys of copper and zinc with the composition, and found that
the action of acids on those alloys exhibited a discontinuity at the
point at which copper and zinc were present in equal proportions.   He therefore suggested the presence of a compound in
the series.

Attempts were next made to isolate definite compounds from
alloys by a process of partial fusion and of mechanical separation
of the solid and liquid phases.   The first extensive experiments
of this kind are due to Crookewit,[8] who examined in this way
many amalgams, and also alloys of copper with tin, lead, zinc,
etc.   This plan was adopted by many other investigators, most
of whom employed amalgams, on account of their low melting-
point and the consequent facility of handling.   The validity of
this method is discussed below (p. 27).

Calvert and Johnson[9] attempted in the same way to isolate
the chemical compounds which they assumed to be present in
alloys, accompanied by an excess of one or the other component,
but their later work[10] opened up a more fruitful field of investigation.   By determining the values of certain physical constants,
such as the thermal and electrical conductivity, hardness, and
specific gravity of a number of alloys in a given series, and observing the manner in which the property selected varied with
the composition of the alloys, they established the fact that discontinuities occur, which were rightly attributed to the presence
of intermetallic compounds.   The same course was followed, with
important results, by Matthiessen and his collaborators, whose
very extensive and accurate determinations of many of the physical properties of alloys placed the subject on a new basis.[11]   The

general conclusions to be drawn from these investigations were summed up by Matthiessen in his " Report on the Chemical Nature of Alloys ".[12] Alloys were here for the first time regarded as solidified solutions, in which definite compounds might or might not be present, and good use was made of the electrical conductivity in distinguishing between the two conditions. Such confusion as is to be found in the interpretation of the curves arises from the presence in many alloys of solid solutions, the nature of which was not at that time understood.

The view that the law of definite proportions was not invariable was advanced by Cooke,[13] as a means of escape from the difficulties presented by the alloys of zinc and antimony, but such an evasion was seen to be unnecessary, and was not adopted by others, although it has been quite recently revived in order to explain certain anomalies in the alloys of bismuth and thallium.[14] This point is discussed below (p. 12). It is clear that only overwhelming evidence would justify any departure from the fundamental principles of the atomic theory, and it will not be generally admitted that obscurities in the behaviour of a few alloys provide a sufficient reason for a reversion to the abandoned view of Berthollet that compounds of indefinite composition may exist.

The modern systematic study of intermetallic compounds begins with the accurate determination of freezing-point curves presenting maxima and discontinuities. The first determinations of this kind were very inaccurate, but they were soon followed by the extremely accurate investigations of Heycock and Neville,[15] who succeeded in establishing the true form of the freezing-point curve for a number of binary metallic systems. The same authors, in their examination of the alloys of aluminium and gold,[16] correlated the thermal and microscopical observations in such a way as to establish the existence of a number of compounds, whilst their classical investigation of the alloys of copper and tin [17] inaugurated the much more difficult and complex study, by thermal and microscopical methods, of that part of the equilibrium diagram which lies below the freezing-point curve.

It is unnecessary to enter in detail into the history of metallography. The principal additions to the list of intermetallic compounds have been made by Tammann and his pupils, in a long list of memoirs from 1903 onwards. For reasons which the author has discussed elsewhere, the experimental methods adopted have been open to criticism, and there is much uncertainty as to many of the proposed formulæ. A gradual revision of the most important systems is now in progress in various laboratories, with the result that, whilst a few systems have been shown to be more simple than had been supposed, others have proved to be unexpectedly complex. The unsatisfactory nature of the available data presents a serious obstacle to any general treatment of the intermetallic compounds, and this difficulty has been felt throughout the preparation of the present monograph. Moreover, the existing material has proved peculiarly intractable to theoretical treatment, and the attempts to form a theory of the constitution of intermetallic compounds, which are briefly discussed in a later section, have been comparatively unsuccessful. The subject offers a promising field of research, and the results are likely to have an important bearing on the further development of atomic and molecular theory.

The number of intermetallic compounds is very large. A list published in 1900[18] included the formulæ of thirty-seven such compounds. In 1909[19] this had grown to 109, whilst a list published in the present year includes 263 formulæ, and the presence of many more compounds has been recognized, although the investigations have not been sufficiently complete for any definite formulæ to be assigned to them. Further research is certain to increase this number very largely.

In the following sections the occurrence of intermetallic compounds as indicated on the equilibrium diagram and determined by the method of thermal analysis is first discussed. This is followed by a short note on the microscopical control of the thermal indications. An account is then given of the methods which have been adopted with the object of isolating intermetallic compounds in a pure condition, and of their assumed occurrence as native minerals. So little success has been met with in this direction, however, that our knowledge of the properties of such compounds is mainly derived from a study of the alloys in which they occur. The succeeding sections are therefore devoted to a consideration of the influence which the presence of intermetallic compounds exerts on some of the most important physical properties of alloys. It is also shown that the systematic investigation of certain properties, especially of the electrical conductivity and the thermo-electric power, affords the most delicate means in a large number of cases of determining whether chemical combination takes place in a given series or not. An account is then given of the scanty data which we possess as to the crystallographic characters of intermetallic compounds, and of the evidence for the existence of compounds in liquid alloys, and the concluding section reviews the theoretical aspect of the subject.*

* The practical methods employed in investigations of this kind are described in the author's "Metallography" (Text-Books of Physical Chemistry, ed. Sir W. Ramsay. 2nd edn. London, 1913).

CHAPTER II.

MOST of the intermetallic compounds added to the literature by recent investigators have been detected in the first instance by the method of thermal analysis. This method depends on the fact that a change from one phase to another, such as the solidification of a liquid, or the polymorphic transformation of a solid, is almost always accompanied by the development or absorption of heat. The exceptional cases in which a change of phase occurs without any appreciable change of thermal energy may be neglected for the moment. Restricting ourselves to the study of binary metallic systems, the method is applied by determining, for a sufficient number of alloys in the series, the temperatures at which such thermal changes take place during heating or cooling. In this way a number of curves are obtained, exhibiting the temperature at which each change takes place as a function of the composition of the alloys, and these curves form the basis of the equilibrium diagram, the construction of which is the first task of the metallographer who undertakes the complete investigation of any system of alloys.

The first curve, and that which is in general most readily determined, is the freezing-point curve or liquidus, representing the variation of the initial freezing-point with the composition. In the early days of thermal analysis, this was the only curve determined,[20] and it was not immediately recognized that its uncontrolled indications might lead to fallacious conclusions. It will, however, be convenient to consider this curve before proceeding to the other parts of the equilibrium diagram.

The liquidus curve of a system of two mutually indifferent metals is usually composed of either one or two branches. If the metals are completely isomorphous, the curve has only a single branch, which may lie entirely between the freezing-points of the two components, or may pass through a maximum or a minimum. If there is a gap in the miscibility of the two metals in the solid state, the curve consists of two branches, which may intersect either at a eutectic point or at a point at which a change of direction occurs without actual reversal.[21] Each branch then represents the separation from the liquid of crystals of a single type, the two types being in this case either the pure metals, or solid solutions of the one metal in the other. The presence of any additional branch of the liquidus indicates that a new type of crystals—a new solid phase—makes its appearance within a certain range of composition. Such a new branch may be due to a polymorphic modification of one of the component metals, or to an intermetallic compound. It is not always easy to distinguish the two cases, and the aid of other methods of investigation must be frequently invoked, but the study of this curve is generally the preliminary to the determination of the formulæ of the intermetallic compounds in any series of alloys. The examination of the liquidus was practised as a means of detecting intermetallic compounds before the principles of thermal analysis were correctly understood, and many erroneous formulæ have in this way found an entrance into the literature. The view was very commonly held that a break in the curve indicated the presence of a compound, the composition of which could be determined by dropping a perpendicular on to the axis of abscissæ. The demonstration of the true nature of eutectics [22] led to a modification of this view as far as eutectic points were concerned, but the recognition of the fact that a discontinuity does not necessarily occur at the composition corresponding with a compound, has only been of slow growth.

The first case which falls to be considered, as exhibiting most clearly the conditions of the problem, is that in which the intermediate branch of the liquidus passes through a maximum temperature. It may occasionally happen that the maximum lies far above the freezing-point of either of the component metals. The interpretation of the curve in such a case is unambiguous. An intermetallic compound must be present, the high freezing-point of which indicates that it is formed from its components with the liberation of a large amount of energy. A familiar example occurs in the amalgams of the alkali metals, which are, within a certain range of composition, relatively very infusible. In every such case it is found that the maximum corresponds with a definite intermetallic compound of great stability. A list of compounds of this class is given below, the freezing-points of the component metals being added for comparison :—

AuMg	1150°	Au	1062°	Mg	650°
Sb_2Mg_3	960°	Sb	629°	,,	
$SnMg_2$	783°	Sn	232°	,,	
Bi_2Mg_3	715°	Bi	270°	,,	
$NaCd_2$	384°	Na	97·5°	Cd	321°
Hg_2Na	360°	,,		Hg	– 38·7°
Hg_2K	270°	K	62°	,,	
Hg_2Cs	208°	Cs	26°	,,	

Similar conspicuous maxima are presented by the tellurides of many metals, compounds which may be considered to form a connecting link between the intermetallic compounds and the sulphides :—

ZnTe	1238°	Zn	419°	Te	451°
CdTe	1050°	Cd	321°	,,	
PbTe	915°	Pb	327°	,,	
HgTe	610°	Hg	– 38·7°	,,	
BiTe	573°	Bi	270°	,,	

When one of the components is a relatively infusible metal, the maximum due to an intermetallic compound, although very strongly marked, may fail to reach the freezing-point of the less fusible component. The following compounds are examples of this class—

PtSn	1280°	Pt	1755°	Sn	232°
$PtSb_2$	1230°	,,		Sb	629°
AgMg	820°	Ag	961°	Mg	650°
AuZn	744°	Au	1062°	Zn	419°
$MgZn_2$	595°	Mg	650°	,,	
Mg_2Pb	551°	,,		Pb	327°
AuSn	418°	Au	1062°	Sn	232°

whilst relative maxima, somewhat less conspicuous, are of very frequent occurrence.

When the liquidus comprises several intermediate branches, it is quite possible for two or more of these to present maxima. This is the case, for example, with the alloys of gold with magnesium, each of the following compounds being represented by a distinct maximum on the curve (fig. 1) :—

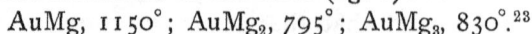

AuMg, 1150°; $AuMg_2$, 795°; $AuMg_3$, 830°.[23]

The exact form of the maximum is discussed later (p. 78) with reference to the condition of intermetallic compounds in the liquid phase. Here it is sufficient to remark that a flattened maximum indicates considerable dissociation of the compound into its components at the melting-point, and that very acute maxima are not met with amongst metallic alloys.

It is unusual for an intermediate branch of the liquidus to pass through a minimum, although minima are of frequent occurrence in the freezing-point curves of isomorphous metals, consisting of a single branch. In the latter case there is no question of a compound. There is no clearly established case of a compound occurring at a minimum.

The case of a branch intermediate in character between the curves presenting a maximum and a minimum respectively, namely, a horizontal branch parallel with the axis of composition, is reported to occur in mixtures of naphthalene and *m*-dinitro-benzene in the neighbourhood of a compound containing 1 mol. of each of the components.[24] Such a curve, which must be regarded as a completely flattened maximum, has not been observed in alloys. Horizontal branches of the liquidus are common, but in every instance they

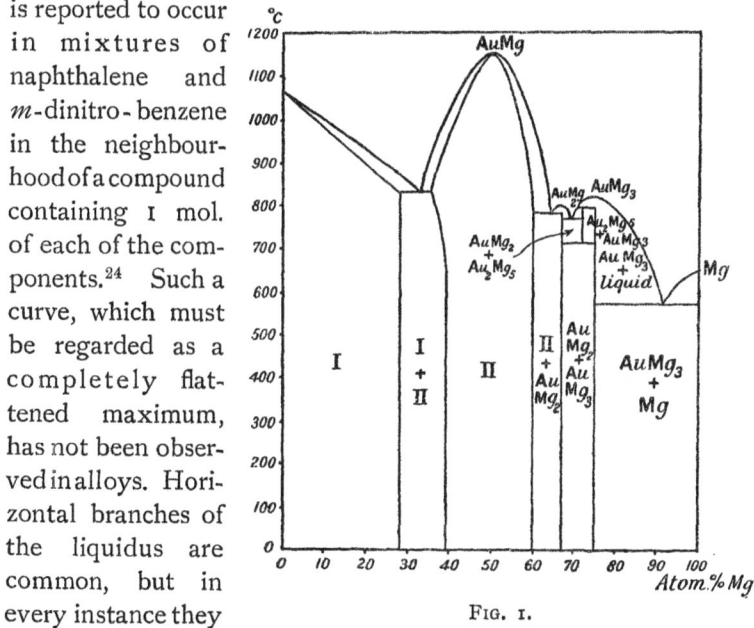

FIG. 1.

have proved to have a different origin, the separation of the molten alloy into two liquid phases.

When the accurate experimental determination of the liquidus has shown that a distinct maximum is present, the inference that an intermetallic compound is formed is a highly probable one, and the composition of the compound is given directly by the position of the maximum. Should the curve be greatly flattened, it may be difficult to determine this position exactly, and the difficulty is increased if the apparent atomic ratio is high or complex. For example, the compound Bi_5Tl_3 is described as presenting a maximum at $227°$,[25] but it is evident that further experiments are necessary before such a formula can be accepted, whilst a maximum described as corresponding with the compound $Cd_{11}K$ [26] is still more open to doubt.

Actually, the establishment of such formulæ by thermal analysis does not depend solely on the determination of the position of the maximum. A great extension was given to the method by Tammann [27] when he proposed a quantitative interpretation of the cooling curves of individual alloys. In addition to the arrests which are employed in the construction of the liquidus, representing the first separation of the solid phase from the molten alloys, the cooling curves may present other arrests, due to the solidification of eutectics, to reactions between the liquid and solid phases, or to transformations within the solid alloys. Any such arrest is limited to a range of composition within which the phases actually concerned are present ; it is absent from all alloys having a composition which lies outside that range. Further, the duration of the arrest, which is proportional to the thermal change in question (all the cooling curves being assumed to be determined under comparable conditions) must be a maximum in that alloy in which the substance undergoing the change occurs in a pure form. For example, a development of heat due to the polymorphic change of a compound must be a maximum for the pure compound, whilst a development of heat due to the solidification of a eutectic must vanish at the composition at which one of the phases composing the eutectic disappears. In a series from which solid solutions are absent, and composed of the two metals A and B, forming a single compound AB with maximum freezing-point, there are two eutectics, the constituents of which are A and AB, and AB and B respectively. In the first half of the series, the eutectic arrest vanishes at the composition AB (50 atomic per cent of B), because alloys containing a larger proportion of B do not contain A as a solid phase, and the same reasoning applies to the second half of the system. On the other hand, if the compound AB undergoes a polymorphic transformation, the arrest due to that transformation is a maximum at 50 atomic per cent. Hence the plotting of the "arrest-times" against the composition gives a valuable means of determining the composition of compounds, always assuming—and this is in practice an important qualification—that the conditions of cooling are such that equilibrium is attained.

Many intermetallic compounds, appearing as maxima on the freezing-point curves, are capable of forming solid solutions with one or both of their components. The capacity for doing so varies within wide limits, being almost absent in such cases as those of the compounds of magnesium with tin, lead or bismuth, whilst it is very strongly marked in the compounds of magnesium with gold. Tammann's method is then applicable with certain restrictions. The arrests due to eutectics do not vanish at the composition of the compounds, but at an earlier point, namely, at the composition of the saturated solid solution containing the compound. Limits are thus determined, within which the composition of the compound must lie, but it is not fixed more precisely. On the other hand, a polymorphic transformation, if such occurs, is still of great value in fixing the point required. Further, as the temperature of transformation is frequently lowered by the presence of a metal in solid solution, the form of the transformation curve may be employed in the same manner as that of the liquidus. All such points as these are best understood after the study of a few actual equilibrium diagrams.

A maximum may occur, when solid solutions are formed, without the presence of a compound. Series of solid solutions with a maximum freezing-point, without distinct evidence of chemical combination, are occasionally met with in organic chemistry, the best-known examples being the mixtures of the isomeric 2 : 4 : 6- and 2 : 3 : 5- tribromotoluenes,[28] and of *d*- and *l*- carvoximes.[29]

An instance of a possibly erroneous interpretation of a maximum occurs in the series lead-thallium. The freezing-point curve of these alloys has a very flat maximum at 380° between 30 and 40 atomic per cent of lead, and this was regarded by Lewkonja[30] as indicating a compound $PbTl_2$, capable of forming solid solutions with both components. On the other hand, Kurnakoff and Pushin[31] concluded that such a compound was absent, although the presence of the maximum was also recognized by them.   The grounds for this conclusion were : the flatness of the maximum, indicating that the compound, if present at all, must be highly dissociated, and the fact that its position was displaced by the addition of tin, the maximum of the curve in the ternary system separating the field of crystallization of tin from that of the solid solution occurring at the ratio Pb : Tl = 1 : 2·5 instead of 1 : 2. This plan of observing the influence of a third metal on the position of the maximum is one of great value in deciding whether a given solid solution contains a compound or not, but it has been resorted to very rarely.   The method of electrical conductivity lends itself particularly to the solution of such doubtful cases, and in this instance the evidence from the conductivity is opposed to the hypothesis of a compound, as neither the conductivity nor the temperature-coefficient exhibits any discontinuity.[32]

A more difficult case of the occurrence of a maximum of doubtful significance presents itself in the alloys of bismuth and thallium.   The freezing-point curve is of peculiar form, having three maxima, at 37·2, 89, and 99 atomic per cent of thallium.[25] The two latter do not correspond with any simple atomic ratio, whilst the first is sufficiently close to 37·5 per cent, the proportion required for a compound Bi_5Tl_3.   It appears, however, that micrographic examination and determinations of the electrical conductivity, temperature-coefficient of resistance, and hardness all agree in fixing the composition of the compound at 36 atomic per cent Tl, which does not correspond with any simple ratio.[14] Kurnakoff and his colleagues therefore conclude that the solid phase in question belongs to the class of " indefinite compounds," the existence of which was maintained by Berthollet as against Proust in the early controversy as to the law of definite proportions.

The system is, however, a very complex one.   Reactions in the solid state point to the formation at low temperatures of another compound, possibly $BiTl_3$.   In view of the great difficulty of obtaining equilibrium in such alloys (the lowest eutectic point being at 186°, whilst the reaction in the solid state occurs below 100°) it is reasonable to assume that the period of annealing was insufficient to ensure equilibrium, and that the alloys employed for the physical determinations contained metastable phases, which would readily lead to errors in the position of the compound.

An intermediate branch of the liquidus which does not present a maximum corresponds with the formation of a new solid phase but not necessarily of a compound. The new phase may be a polymorphic modification of the solid phase separating along the next higher branch of the curve, or it may contain a new compound. The two cases are not always readily distinguishable, but an application of the method of plotting arrest-times is often successful in deciding between them. For example, the liquidus curve in Fig. 2 exhibits discontinuities at *r* and *p* in addition

to the eutectic point, that is, the liquidus has two intermediate branches. We may suppose that the heat-change taking place at the temperature of the line *pq* is found to be a maximum at the composition AB_2, as shown by the inverted dotted curve. It may then be attributed to the formation of a phase having that composition

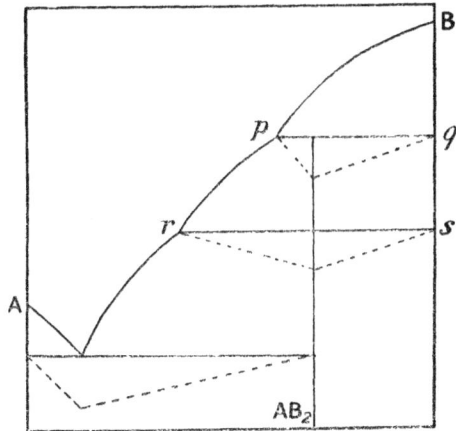

FIG. 2.

from crystals of solid B and the liquid phase during cooling, or to the corresponding decomposition on heating, represented by the equation.

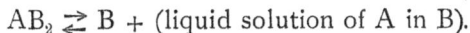

$$AB_2 \rightleftarrows B + \text{(liquid solution of A in B)}.$$

The second break in the liquidus of Fig. 2 is exactly similar in aspect to the first. An application of the method of "arrest-times" to the alloys between *r* and *s* shows that the maximum heat-change occurring along the line *rs* is found at the same composition, AB_2, as shown by the second dotted curve. The arrest must therefore be due to a polymorphic transformation of that compound. Had it been due to a further reaction between the solid phase and the liquid, the maximum would not have been found at the composition AB_2, but at some point further to the left, such as AB or A_2B_3.

If the compound is capable of entering into solid solution with one of its components, such a diagram as that in Fig. 2 is modified, as the lines *pq* and *rs* are not continuous as shown in the diagram. The principle of thermal analysis, however, remains the same, and the maximum heat-change occurs at the composition corresponding with that of the compound, if the conditions are such as to favour the attainment of equilibrium. In actual practice, the slowness of a reaction between two solid phases and a liquid, such as that on the line *pq*, tends to prevent the complete attainment of equilibrium, and the maximum heat-change appears to be more or less displaced. This error may be avoided or corrected by means which are described in the text-books of metallography, or in the original papers of Tammann.[27] Microscopical examination furnishes a means of determining the extent of the deviation from equilibrium, and of computing the true composition of the compound, whilst the examination of fully annealed specimens furnishes the most satisfactory means of delimiting the regions of stability of the respective phases.

The application of these principles may be best illustrated by a concrete example. The alloys of magnesium and silver have been subjected to thermal analysis with great care, and apparently with satisfactory experimental precautions.[38] The equilibrium diagram constructed from the thermal observations is given in the upper part of Fig. 3. The liquidus presents a single maximum, and also a discontinuity of the kind just described. The alloys were not perfectly homogeneous after annealing, and it is possible that under conditions more nearly approaching equilibrium, the distance between the liquidus and solidus curves on the right-hand portion of the diagram would be still further reduced, whilst the vertical lines bounding the regions of solid solutions I and II would be slightly displaced. Solid solutions are not formed to any considerable extent in alloys containing less than 25 atomic per cent of silver.

The discontinuity in the liquidus curve does not occur at the composition of the compound $AgMg_3$, but somewhat to the left of it, namely, at 23 atomic per cent Ag. The position of the vertical line at 25 per cent indicates that the thermal effect of the reaction

solid solution I + liquid → $AgMg_3$

should be a maximum at that composition, and also that the thermal effect due to the solidification of the eutectic should vanish at that point. The conclusions from the thermal results have been fully confirmed by the study of the mechanical and electrical properties of the alloys (see below, p. 43).

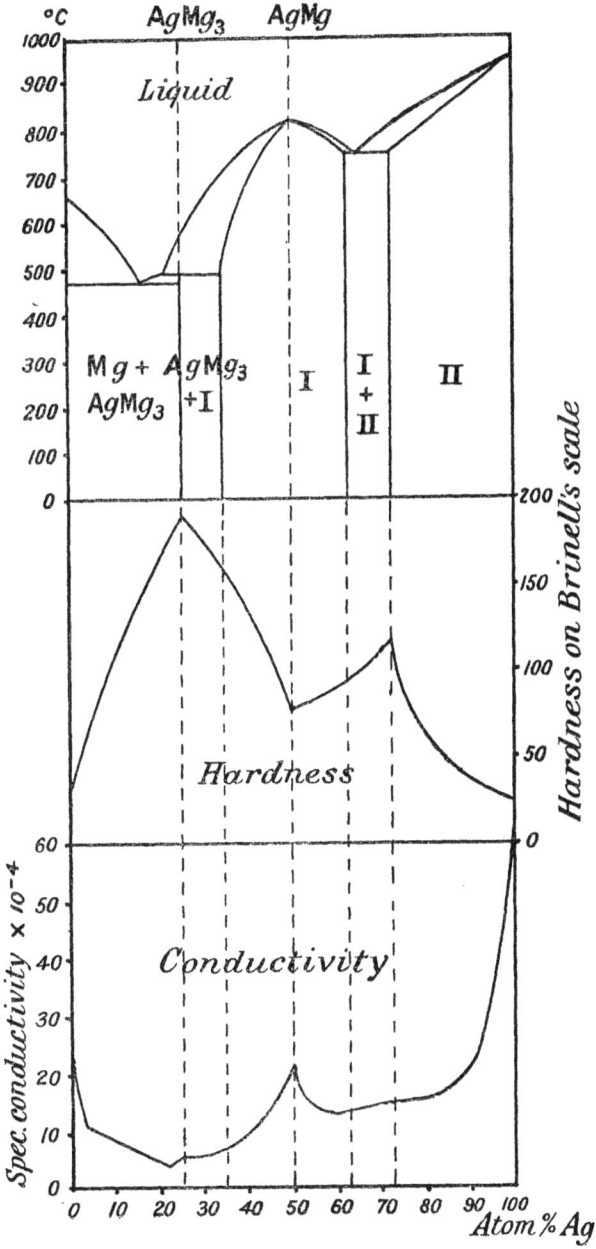

Fig. 3.

A compound which is capable of entering into solid solution in all proportions with both of its components may give rise to a peculiar modification of the form of the liquidus curve, an example of which is seen in the alloys of magnesium and cadmium (Fig. 7). The compound MgCd forms a continuous series of solid solutions with both magnesium and cadmium. There is thus a point of inflexion in the liquidus curve at 50 atomic per cent, and the solidus, which at this point coincides with the liquidus, diverges from it in both directions, only converging again towards the pure metals at the limits of the diagram. It is probable that at the composition of the compound both liquidus and solidus have a horizontal tangent, but the curves have not yet been determined experimentally with sufficient accuracy to decide this point. Further evidence of the presence of the compound is afforded by a decomposition which takes place at a lower temperature. The crystals, which are homogeneous throughout the series at temperatures immediately below the solidus, undergo resolution on further cooling, and the curve representing the change passes through a maximum temperature at the composition of the compound MgCd. Repeated reference is made in the sequel to this remarkable system.

Different views have been held by investigators as to the interpretation of intermediate branches of the liquidus. In exceptional cases such a branch may be clearly attributed to the existence of a polymorphic modification of one of the components, but in some other cases the interpretation is less obvious in the absence of a maximum freezing-point. On the one hand, Tammann and his school have assumed that an intermediate type of crystal must contain an intermetallic compound, and an attempt has been made to assign a formula in every such case. On the other hand, Bancroft and his school have inclined to reject the assumption of a compound unless clearly indicated by a maximum or in some other way. On this view, two metals may form, not merely two series of solid solutions separated by a gap, as shown on theoretical grounds by Roozeboom, but any number of such series, separated by a corresponding number of gaps. For example, in a study of the alloys of copper and zinc, Shepherd [34] has shown that these metals form no less than six different solid phases α, β, γ, δ, ϵ, and η, each of which is represented by a distinct branch of the liquidus, and he does not regard any of these phases as corresponding with a definite compound. The question evidently cannot be decided from a thermal study alone, as in this example there do not appear to be any polymorphic changes which would be of assistance in applying the method. There is, however, evidence of other kinds to show that the γ-phase contains a definite compound, Cu_2Zn_3, whilst the presence of a second compound, CuZn, in the β-phase is very probable, and a more detailed study of the physical properties of the series may reveal the presence of still further definite compounds. Another example is that of the alloys of antimony and tin.[35] The liquidus is composed of four branches, falling from the freezing-point of antimony to that of tin, with scarcely perceptible changes of direction at their junctions. The solid phases may be considered either as four series of solid solutions, separated by gaps, or the two intermediate types may be regarded as containing the compounds Sb_2Sn_3 and SbSn respectively. The electrical conductivity of the alloys[36] does in fact point to the presence of these compounds in the two solid phases.

A decision on this question does not affect the form of the equilibrium diagram in any way. The two opposing views refer only to the molecular condition of the solid solutions concerned. On the one view, the metals are present, at least in a large part, in the form of compound molecules, on the other, they are free. The physical methods for the investigation of solid phases, which are discussed later, are capable of furnishing decisive evidence, and the tendency of the limited number of suitable investigations hitherto available is to show that compounds, sometimes in a largely dissociated condition, are present in the solid phases under discussion.

The next class of intermetallic compounds comprises those which are formed from their components in the solid state, and are not capable of existence in contact with the liquid phase. The detection of such compounds by the method of thermal analysis necessitates careful manipulation and a delicate method of recording arrests on the heating or cooling curves. Consequently, although there is little doubt that the case is one of fairly frequent occurrence, the experimental evidence is not often sufficiently complete to establish the conditions with certainty.

The simplest case is that in which the two metals crystallize in a pure state, that is, without forming solid solutions to an appreciable extent, and the solid components then unite to form a compound at a temperature below the eutectic point. As the reacting phases are the two metals, the temperature of combination remains constant throughout the series, and the reaction

$$mA + nB = A_mB_n$$

is represented on the equilibrium diagram by a horizontal line (Fig. 4). Such a case is stated to occur in the alloys of lead and antimony. The eutectic point is at 248·5°, and an arrest has been observed on the cooling curves only slightly below this, at 244·8° [37]. The magnitude of the heat-change at this point is small, and appears to attain a maximum in the neighbourhood of 60 atomic per cent of antimony, pointing to the formation of such a compound as Pb_2Sb_3. This explanation is provisionally adopted by Guertler,[38] but demands further verification. The difficulty of obtaining equilibrium in such alloys is particularly great, as the supposed

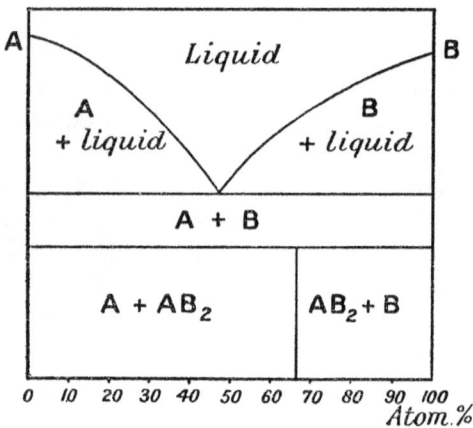

reaction is one between two solid phases at a temperature so low that the molecular mobility is extremely small. The evidence is entirely derived from the thermal measurements, as it has not been found possible to recognize the formation of a new micrographic constituent in the annealed alloys. Pre-

FIG. 4.

cisely similar conditions have been suspected in the alloys of lead and tin, a development of heat at 155° during cooling being attributed to the formation of a compound, Sn_3Pb_4.[39] The evidence here is even less satisfactory, and the thermal effect observed has been attributed to polymorphism,[40] and to the undercooling of solid solutions.[41] The same remarks apply to the alloys of tin and bismuth, in which the development of heat occurs at 95°, tin and zinc (161°), tin and cadmium (130°), and tin and mercury (− 38·6).[38]

On the whole, the hypothesis of the formation of a compound in these alloys must be regarded as an improbable one, in view of the facts that microscopical evidence is lacking, that the chemical relations of the metals concerned are not such as to suggest the probability of combination, and that polymorphism or undercooling is quite capable of giving rise to the effects observed.

The difficulty of obtaining satisfactory evidence from a microscopical examination in such a case is obvious from Fig. 4. The two metals A and B are assumed to crystallize in the pure state. Immediately below the eutectic temperature, therefore, the solid alloys consist of the two solid phases A and B, intermixed in the form of a eutectic, in which are embedded crystals of the one or other metal, according to the composition. At a lower temperature, a new solid phase is formed by the reaction—

$$A + 2B = AB_2$$

this reaction proceeding at constant temperature. The reacting substances, however, are solid and in a dispersed form, and all alloys except that which contains 33·3 atomic per cent of A have an excess of one or other constituent above that proportion which is required for the reaction. These conditions tend to hinder the attainment of equilibrium, and in the case of metals of low melting-point, the molecular mobility is so small that annealing for many weeks may be necessary in order that the formation of the new phase may proceed to an appreciable extent. None of the systems mentioned above have yet been studied from this point of view with sufficient accuracy.

A complex case, which is, however, better supported by microscopical evidence, is that of the alloys of nickel and tin,[42] in which a compound, Ni_3Sn, stable above 855°, reacts with another solid phase to form a new compound, which is either Ni_4Sn or Ni_6Sn. At a somewhat lower temperature (837°) alloys containing a larger proportion of tin undergo decomposition, the same compound being formed by dissociation of the compound Ni_3Sn:—

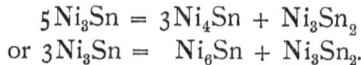

$$5Ni_3Sn = 3Ni_4Sn + Ni_3Sn_2$$
$$\text{or } 3Ni_3Sn = Ni_6Sn + Ni_3Sn_2.$$

Even this statement must be regarded as hypothetical, although the formation of a new solid phase is certain.

It is possible for the range of stable existence of a compound to be limited in temperature both upwards and downwards. An example presents itself in the alloys of gold and magnesium, which were independently investigated by Urazoff[23] and Vogel.[43] Certain differences between these two authors were subsequently reconciled in a joint paper.[44] The probable equilibrium diagram is shown in Fig. 1. The compound Au_2Mg_5 has no definite melting-point, but decomposes at 796° into crystals of the compound $AuMg_3$ and a liquid alloy. On the other hand, the solid compound, which does not form solid solutions with its components, decomposes on cooling past 716°, according to the equation

$$Au_2Mg_5 \rightleftarrows AuMg_2 + AuMg_3,$$

the thermal and microscopical evidence combining to confirm this conclusion.

The case of a compound which, whilst stable at high temperatures, undergoes decomposition with falling temperature, presents certain features of interest. A possible example, in which the reaction takes place without the formation of solid solutions, has been mentioned above (Ni_3Sn, p. 19), but the case is more frequent in systems which include solid solutions.

It has recently been shown[45] that a great similarity exists between the three systems, copper-zinc, silver-zinc, and silver-cadmium, especially as regards alloys containing from 0 to 60 atomic per cent of the more fusible metal. The three partial diagrams are reproduced in Fig. 5. The form of the liquidus curves between these limits indicates that two series of solid solutions, designated a and β respectively, crystallize from the molten alloys. The third solid phase, γ, is in each case a definite compound, Cu_2Zn_3, Ag_2Zn_3, or Ag_2Cd_3. The a-solution has a considerable range of composition, extending from 0 to between 30 and 40 atomic per cent at the ordinary temperature. The β-solution is in each case remarkable in that its range of composition narrows on both sides with falling temperature, ultimately vanishing at a eutectoid point. There is no conclusive evidence that a compound, CuZn, AgZn, or AgCd is actually present in the β-solution, but the position of this branch of the liquidus in the diagram is strongly suggestive of its presence, and the electrical evidence, discussed below, points to a similar conclusion.

The formation of a compound during the cooling of an alloy may, in the absence of nuclei of that compound, be suppressed by undercooling. This is most likely to occur when the compound is formed by a reaction between a solid and a liquid phase, or between two solid phases. It has, however, been observed also in the solidification of liquid alloys. Thus, in the system cadmium-antimony [46] the freezing-point curve has a maximum at 455°, corresponding with the solidification of the stable compound CdSb. Between this point and the freezing-point of antimony there is a eutectic point at 445°. In alloys containing more cadmium, a reaction takes place at 410° :—

$$CdSb + liquid \to Cd_3Sb_2.$$

Freezing only takes place in this way, however, if the alloy is inoculated with the solid compound CdSb. Failing inoculation, nuclei of this compound do not appear as the temperature falls, and a different (metastable) freezing-point curve is obtained,

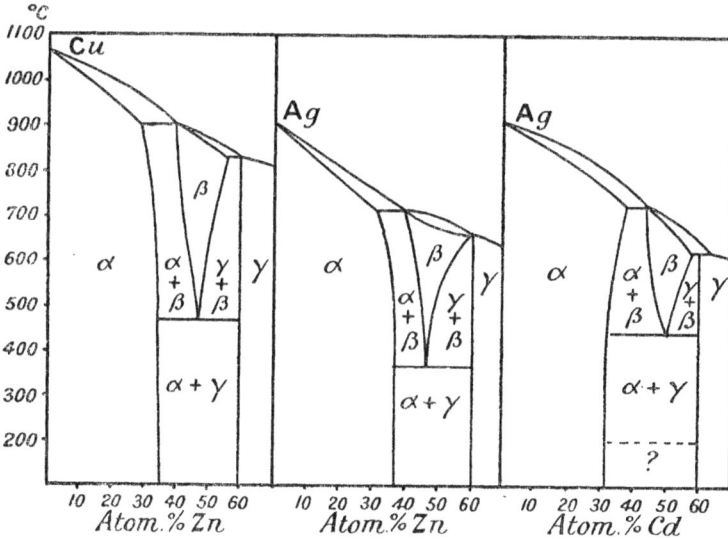

FIG. 5.

having a maximum at 423°, corresponding with the compound Cd_3Sb_2, and a eutectic point at 402°, at which the two solid phases Cd_3Sb_2 and Sb are in metastable equilibrium. If alloys containing from 40 to 50 atomic per cent Sb are cooled in this way a rearrangement takes place in the solid state, and the passage to the stable condition, in accordance with the equation

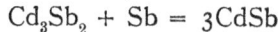

$$Cd_3Sb_2 + Sb = 3CdSb$$

takes place more or less completely, with the development of a considerable amount of heat.

The alloys of zinc and antimony behave in a similar manner, but in this case the maximum on the curve of the stable system represents the compound Zn_3Sb_2.[47]   It is the compound ZnSb which is formed under stable conditions by a reaction between Zn_3Sb_2 and antimony, and does not make its appearance in the absence of inoculation.   A metastable equilibrium between Zn_3Sb_2 and antimony is thus obtained, and a eutectic point, 23° below the normal eutectic point, makes its appearance.

The two systems just described are closely analogous, and it is interesting to observe that very similar conditions have recently been found to present themselves in the related system cadmium-arsenic.[48]   There are two maxima on the freezing-point curve, at 721° and 621°, corresponding with the compounds Cd_3As_2 and $CdAs_2$ respectively.   In the absence of inoculation with crystals of the latter compound, its formation from the liquid is suppressed by undercooling, and the descending branch of the curve representing the freezing of Cd_3As_2 may be prolonged 92° into the metastable region, ending in a eutectic point Cd_3As_2 – As.   At this point the reaction

$$Cd_3As_2 + 4As = 3CdAs_2$$

takes place with development of heat, the system differing in this respect from the two systems described above, in which reversion to the stable condition does not take place until after complete solidification.

This liability to undercooling may persist in the presence of a third metal.   Thus, in the examination of the ternary alloys of cadmium, copper, and antimony,[49] it was observed that the formation of the compound CdSb depended on inoculation, failing which a metastable equilibrium was obtained, in which the solid phases, instead of being Cu_2Sb and CdSb, were Cu_2Sb, Sb, and Cd_3Sb_2.   Then, at a temperature (262°) much below the solidus, a sudden development of heat was observed, due to the combination between two solid phases and reversion to the stable system.

CHAPTER III.

MICROSCOPIC STRUCTURE.

IN any complete investigation of a series of alloys, the microscopical examination should go hand in hand with the thermal study. The examination of metals under the microscope at temperatures other than atmospheric has proved impracticable, and etching at high temperatures has found only a very limited application. The metallographer is therefore restricted in practice to the examination of alloys, the structure of which has been developed by suitable means at the ordinary temperature. It follows that a comparison of the microscopic structures of a series of slowly cooled alloys reveals the solid phases which are in equilibrium with one another at about 16°. In other words, it determines the intercepts of the vertical and inclined lines of the equilibrium diagram on a horizontal line drawn across the diagram at 16°. It is possible, however, to obtain much more information than this with the aid of the microscope. By the device of rapidly quenching selected alloys from known temperatures, and developing the structure of the quenched specimen by etching, it is possible to determine the phases which were present in the alloy at the temperature of quenching, the assumption being made that the rate of cooling was sufficient to suppress completely all changes taking place at a lower temperature. It often happens that this condition is not strictly fulfilled, but the appearance of the microscopic structure usually reveals to the practised observer the nature of the change that has taken place, and it is possible to allow for it, more or less accurately, in drawing conclusions from the examination.

The simplest example is that of a binary metallic system in which a single compound occurs, and solid solutions are not formed to an appreciable extent. Starting from the pure metal A, the addition of small quantities of B leads to the appearance of a eutectic in the alloys, which increases with increasing proportions of B until it composes the whole of the alloy. When

23

B is increased beyond the eutectic proportion, crystallites of the compound A_mB_n make their appearance in increasing quantity. When the exact composition A_mB_n is reached, the alloy is homogeneous. When the proportion of B is increased still further, a similar series of changes in structure is passed through in the reverse order, the quantity of crystallites of the compound diminishing until the second eutectic alloy is reached. The determination of the composition of the compound is a very accurate one, as the presence of a very small quantity of the first or second eutectic by means of the microscope is easy.

When the compound is one which has been directly deposited from the liquid alloy at a maximum freezing-point during cooling, the application of the above method is comparatively simple. It is liable to greater error if, on the other hand, the compound has been formed by a reaction between a solid and a liquid phase. On account of the slowness of diffusion in solids, the reaction is very apt to remain incomplete under ordinary conditions of cooling, and the solid phase of later formation coats the crystallites of the earlier phase, and hinders further reaction with the surrounding material. Hence such alloys are frequently seen when cold to consist of three solid phases. As such a condition is incompatible with equilibrium in a binary system (except at a triple point), warning is at once given by such a structure that the alloy is in an unstable condition. Equilibrium is most readily attained by cooling rapidly in the first instance to ensure a fine-grained structure, and then again heating the alloy and maintaining it for a sufficient time in the neighbourhood of the temperature of reaction. Diffusion is thus facilitated. This treatment is often desirable, as the case just discussed is of very frequent occurrence.

The conditions are again altered by the formation of solid solutions of the compound with one or both of its components. Reverting to the case first discussed, the structure of the alloys becomes homogeneous before the composition A_mB_n is reached, or remains so until a further quantity of B is present, or the homogeneous region may extend in both directions. It is now only possible to conclude as to the composition of the intermetallic compound that it lies somewhere within the limits of homogeneous structure, and examination by means of the microscope does not yield further information.

When the diagram is more complex, the rule remains that a line drawn horizontally across the diagram must pass alternately through fields representing homogeneous and heterogeneous alloys. When the miscibility in the solid state is very small, the breadth of a homogeneous region may be inappreciable, but it must nevertheless exist. The gradual disappearance of one eutectic and appearance of another as the proportion of the second component is increased is usually easy to observe, and affords an accurate means of fixing the limits of the homogeneous region. Where eutectics are absent, as when alloys separate into two constituents on cooling, similar considerations apply, the second solid solution playing the same part in the micrographic examination as a eutectic.

The brittleness of many intermetallic compounds is a source of difficulty in the practical application of the micrographic method. The grinding of highly brittle alloys results in the removal of small chips, and the cavities thus formed may present the appearance of a second constituent, or at least render it difficult to determine whether the structure is really homogeneous or heterogeneous. This difficulty can only be overcome by careful grinding, using very light pressures. A smooth carborundum block may usefully replace the ordinary emery papers, and repeated alternations of polishing and etching will often lead to a successful result.

When a homogeneous region in the equilibrium diagram is found to exist, the question whether an intermetallic compound is present must remain an open one, as far as the micrographic examination is concerned. It is then necessary to resort to a study of one of the electrical properties in order to arrive at a decision.

The characteristic colour of some intermetallic compounds should be noticed. The γ-phases of some of the copper alloys are bluish-white, and many of the harder compounds of the heavy metals are grey. Two compounds are remarkable for their violet or purple colour, namely, Cu_2Sb, which forms a characteristic eutectic with antimony, and Al_2Au, known as Roberts-Austen's purple alloy.

CHAPTER IV.

THE ISOLATION OF INTERMETALLIC COMPOUNDS.

IN other branches of inorganic as well as of organic chemistry, the isolation of a compound in a pure condition is usually a necessary preliminary to its study and description. Mixtures containing the compound are subjected to the physical processes of crystallization, distillation, etc., by means of which the accompanying substances are in turn removed, until the constancy of properties of the resulting product under further treatment indicates that a definite compound has been obtained. Such physical processes are supplemented, or even entirely replaced, by treatment with chemical reagents, which are so chosen as to convert the accompanying impurities into easily removable substances, whilst leaving the desired product unattacked.

It is natural that similar methods should be applied to metallic alloys, and the number of investigations conducted on this plan is in fact very large. For reasons which will now be examined, the results have been unsatisfactory and frequently misleading, and chemical literature has been burdened with an extremely large number of formulæ and descriptions of supposed compounds, the majority of which are probably erroneous. Before the doctrine of phases had been applied to alloys, the improbability that intermetallic compounds could be isolated by such means was not apparent, but it is unfortunate that many memoirs continue to be published, especially in French journals, in which the defective methods are employed, in spite of their now fully demonstrated unfitness for the purpose.

Amongst physical methods, those of fractional crystallization and fractional distillation alone suggest any probability of success when applied to the isolation of intermetallic compounds from alloys. Crystallization from an indifferent solvent is inapplicable, the only available solvents being other metals, which form alloys with the material under examination. The method

26

of fractional crystallization is attended by certain experimental difficulties, of which the most serious is the impossibility of removing the adhering mother-liquor completely from the crystals, washing being impracticable on account of the immiscibility of molten metals with other liquids. Nevertheless, the method has been frequently employed, some attempt being generally made to reduce the adherence of the mother-liquor to a minimum or to allow for its presence by means of some special device. Its application offers the greatest simplicity in the case of amalgams, owing to the fact that mercury is liquid at the ordinary temperature, and the validity of the method is most conveniently examined in this case. A careful study of amalgams, with the object of determining whether definite chemical compounds were formed, is due to Kerp, [50] [51] whose results served as the basis for the formulæ adopted in textbooks until very recently, when they were corrected by the more trustworthy method of thermal analysis.

Some early analyses of the crystals obtained by separating the solid and liquid portions of sodium and potassium amalgams gave the formulæ $NaHg_6$ and KHg_{12}, with comparatively little variation of composition. Apparently definite products of this kind had been previously obtained by other workers. The more detailed investigations of Kerp showed that over the range of temperature from 0° to 100° the solid phase in equilibrium with liquid sodium amalgam was of constant composition, containing 2·13 per cent of sodium, whilst the formula $NaHg_5$ requires 2·24 per cent. The crystals were separated by draining on a Gooch crucible, kept at the temperature of the experiment, using a vacuum pump. In a similar manner, crystals corresponding with the following definite formulæ were obtained :—

$$KHg_{12}, \ BaHg_{12}, \ LiHg_5, \ RbHg_{12}.$$

Kerp's later experiments also led to the conclusion that a second sodium amalgam, $NaHg_6$, was the stable solid phase at temperatures below 40°, whilst several other potassium amalgams were recognized. The following compounds were also described :—

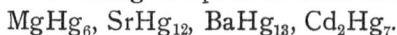

$$MgHg_6, \ SrHg_{12}, \ BaHg_{13}, \ Cd_2Hg_7.$$

A comparison with the results of thermal analysis renders it evident that the crystals were not completely free from mother-liquor, and that the proportion of mercury is consequently in all cases too high.

The manner in which mercury adheres to the crystals in such experiments as these is very remarkable. Crystals which have been drained by means of the filter-pump and pressed down by a pestle until mercury ceases to escape may appear to be entirely free from adhering liquid, whilst the application of considerable pressure, as by wrapping in chamois leather and squeezing, is followed by the expulsion of a further quantity of liquid. Such observations even led to the assumption [52] that the composition of the solid phase is altered by pressure, but it is obvious that the very moderate pressures employed could have no such effect, and that the liquid is only held mechanically by the solid. Kerp observed that some of the solid amalgams formed loose masses of crystals, traversed by capillary cavities, and that such a mass, when immersed in mercury or liquid amalgam, soaked it up like a sponge, the liquid thus absorbed being again expelled on the application of sufficient pressure.

The following method, which has been applied to compounds of the alkali metals, offers a closer analogy with the method of precipitation so commonly employed in other departments of chemistry.[53] The alkali metal and the other metal with which combination is required are melted separately under liquid paraffin, and mixed at a temperature below the melting-point of the compound. A crystalline precipitate is obtained, and in some cases, such as that of sodium bismuthide, this precipitate is insoluble in an excess of sodium. By employing the theoretical quantities of the two metals, the compound is obtained in an apparently pure condition. The crystals are then merely cooled and washed with benzene or light petroleum to remove paraffin, but if an excess of alkali metal has been employed, this must be removed by means of liquid ammonia. The compounds, which are frequently highly reactive, are then dried in an atmosphere of nitrogen.

Vournasos obtained by these means the following compounds, the existence of which has been independently established by thermal analysis :—

$$Na_3Bi, \text{ small leaflets} \quad . \text{ m.p. } 776°$$
$$K_3Bi, \quad \text{minute crystals} \quad ,, \quad 671°$$
$$Na_2Pb, \quad . \quad . \quad . \quad ,, \quad 405°$$
$$Na_2Sn, \quad . \quad . \quad . \quad ,, \quad 477°$$

If the second metal, for example bismuth or tin, be present in excess, an alloy is formed, from which the compound is only isolated with great difficulty. Considerable heat is evolved during the formation of the bismuthides, but the compounds are nevertheless highly reactive, being spontaneously inflammable in dry air.

Liquid ammonia had been previously employed for removing an excess of an alkali metal from alloys, and the following compounds had been isolated by its means [54] :—

$$Na_3As, Na_3Sb, Na_3Bi, and Na_4Sn.$$

Lithium and antimony were even found to react at a low temperature when brought into contact below the surface of liquid ammonia, forming the compound Li_3Sb.[55]

The practical difficulties of isolation are considerably increased if the temperature at which solid and liquid phases have to be separated be above the atmospheric temperature, as is the case with all but a very few alloys. The liquid may be expelled by means of a centrifugal machine,[56] but its removal is always imperfect. An ingenious device has been employed by several investigators,[57] [58] consisting in the addition to the system of a small known quantity of a foreign substance which dissolves in the liquid, but does not enter into the composition of the solid phase. When the crystals are subsequently analysed, a determination of the proportion of added material enables the quantity of the adhering mother-liquor to be calculated. This method has not often been employed. An instance of its use is found in a study of the tin amalgams.[59] Definite quantities of cadmium were added to the amalgams, and the crystals and the liquid were analysed after separation. On the assumption that the whole of the cadmium remained dissolved in the liquid, the quantity of mother-liquor adhering to the crystals could be calculated, and the composition of the solid phase was thus determined. In another case, a small quantity of silver was added to alloys of tin and antimony, in order to determine the quantity of mother-liquor retained by the solid phase.[57]

In the absence of any such special device, there are no means of determining how much of the mother-liquor has been mechanically retained, and the clear and glistening surface of the crystals is no sufficient guarantee of their chemical homogeneity. A few concrete examples may illustrate this point. When copper and aluminium are melted together in about equal proportions by weight, and the mother-liquor is poured off after a portion of the alloy has solidified, large, glistening, silver-white crystals, several centimetres in length, may be obtained.[60]   An analysis of these crystals gives figures which correspond very closely with the formula Cu_4Al_9.   The thermal analysis of the alloys of copper and aluminium has, however, made it clear that no such compound is formed, and that an alloy of such a composition would be made up of δ-crystals and eutectic.   The δ-constituent is, in all probability, a definite compound, $CuAl_2$, and the higher proportion of aluminium found in the experiments just quoted is due to the presence of a considerable quantity of liquid eutectic, relatively richer in aluminium than the crystals, adhering to their surface at the moment of removing the mother-liquor by decantation.   The experiment was repeated by Carpenter and Edwards, who also found it impossible to separate the δ-crystals completely from the eutectic.[61]

The general remarks made above (p. 7) in connexion with the principles of thermal analysis will suffice to make it clear that no reliance can be placed on formulæ obtained by such a method as that just described.   Admitting, even, that it is possible so far to overcome the experimental difficulties as to obtain a correct analysis of the crystals, the values obtained merely determine the composition of the solid phase in equilibrium with the liquid at the temperature of experiment.   Such a phase may be a definite compound, but it may also, with somewhat greater probability, be a solid solution, which may or may not contain a compound. The validity of the formula calculated from the results of analysis is therefore dependent on the absence of solid solutions, a condition which can only be determined by thermal analysis or by an electrical method, either of which yields the required result without the necessity of isolating the solid phase.

The process of distillation has been applied in cases in which one of the component metals of the alloy is volatile at a moderate temperature.   It is rendered more generally applicable by the employment of reduced pressure.   The isolation of a compound AuCd by the distillation of an excess of cadmium from alloys of cadmium and gold [62] is discussed in connexion with the existence of compounds in the liquid state (p. 85).   The compounds AuZn and Cu_3Sb, the existence of which is proved by other evidence, have similarly been obtained in a condition approaching to purity by removal of an excess of zinc and antimony respectively by distillation at 1100° under a pressure of 1 mm.[63] The crystalline compounds Cd_2Pt and Mg_2Pt were also obtained by removing the excess of volatile metal by distillation.[64]

The treatment of alloys with chemical reagents for the purpose of removing an excess of one component, leaving a residue of the intermetallic compound, has been adopted very frequently, and has obtained an undeserved confidence, which is only gradually disappearing before criticisms grounded on the doctrine of phases. A reagent being added which dissolves one of the component metals without acting on the other, the reaction frequently comes to a standstill when a part only of the soluble metal has been removed, the assumption being made that the undissolved portion is in a state of chemical combination with the second metal. In order to avoid the objection that two or more solid phases may be intimately entangled, the alloy may be powdered before being submitted to the action of the reagent, and whenever possible the action of two or more distinct reagents is examined. When two chemically dissimilar reagents attack the alloys, leaving a residue of the same constant composition, the inference that a definite compound has been obtained is considerably strengthened.

The validity of the method is dependent on the compound sought for having a greater resistance towards the reagent employed than the component present in excess in the alloy taken for the experiment.

The practical difficulties of the examination of residues, and the errors of interpretation to which they may easily give rise, are well illustrated by the example of the copper silicides. Earlier experiments[65] had led to the conclusion that a single stable compound, Cu_4Si, existed in this series. In a detailed investigation, Philips[66] found it impossible to obtain a definite silicide from alloys containing an excess of copper, as all reagents capable of dissolving the free copper attacked the silicide to a greater or less extent, whilst a concentrated solution of potassium cyanide even dissolved the alloy completely. On the other hand, alloys containing an excess of silicon, if attacked by means of potassium hydroxide, left a residue which appeared to be of a definite character, having the formula Cu_7Si_2. The same residue was obtained when the reagent used was potassium carbonate. An examination by the thermal method, assisted by microscopical examination, had previously indicated the presence of Cu_3Si, fairly distinctly marked as a maximum on the freezing-point curve.[67]   The only other silicide recognized in this thermal study was considered to have the composition Cu_9Si_4, a formula in itself improbable, and supported by such insufficient evidence that it may be unhesitatingly rejected.   A second silicide probably exists, but it is impossible to deduce its formula from these observations.   On the silicon side of the first eutectic point, however, the freezing-point curve has been more satisfactorily determined, and there is no indication whatever of the presence of a compound Cu_7Si_2.   It thus appears that the method of the analysis of residues tends to indicate too low a proportion of silicon in the solid phase.

The nature of the error involved has been discussed by Guertler,[68] who bases his conclusions on previous observations in the study of metallic silicides.   An alloy consisting of a silicide together with free silicon is attacked by alkali hydroxides, the silicon being removed as silicate, but the action is never confined to that part of the silicon which is present in the free state.   The silicide is also attacked superficially, although less rapidly, with the result that its particles become coated with a layer of metal, which hinders further solution and produces a fictitious appearance of equilibrium.   On analysis, the apparently unattacked residue is found to contain a proportion of silicon which is appreciably less than that which was actually present in combination before the attack.   This behaviour has been observed in the alloys of iron [69] and of nickel [70] with silicon, and accounts fully for the anomalous results obtained by Philips.

An effect of this kind is by no means confined to silicides, but may present itself in the analysis of metallic residues from other alloys, the undissolved portion becoming coated with an impervious layer of the insoluble component, the proportion of which is consequently found on analysis to be too high.

Moreover, the fundamental assumption which underlies the method of the examination of residues, namely, that an intermetallic compound is less readily attacked by reagents than the more reactive of its components, is not always justified.   Guertler mentions the following instances of intermetallic compounds which are more reactive than their components :—

Compounds of magnesium, including Mg_2Pb, Mg_2Sn, and

Mg_3Tl_2 fall to powder in air, oxidation taking place. Sodium antimonide and bismuthide, Na_3Sb and Na_3Bi, are spontaneously inflammable when rubbed. The compound Bi_5Tl_3 becomes coated in moist air with a white powder, whilst its components are unattacked. The alloys of platinum and lead are more readily oxidized than pure lead. Some of the copper silicides are decomposed by boiling water. To these may be added the pyrophoric alloys of iron and other metals with cerium. Several aluminium alloys are liable to spontaneous disintegration in air, but it is not quite certain that this property is due to the intermetallic compounds present, as the alloys as usually prepared may possibly contain readily decomposable carbides.

CHAPTER V.

NATIVE INTERMETALLIC COMPOUNDS.

COMPARATIVELY few of the metals are found in the earth in the native state, such occurrences being almost confined to the iron group (iron, cobalt, and nickel), the metals, mercury, copper, silver, and gold, the group of the platinum metals, and the semi-metals arsenic, antimony, bismuth, and tellurium. It is therefore not surprising that few intermetallic compounds are known to occur as minerals, and that the majority of those which are known are compounds with the semi-metals, and therefore approach in chemical character the sulphides and other typical mineral compounds.

That a mineral, composed of two or more metals, has a definite crystalline form and an approximately uniform composition, does not prove it to be a definite compound. In many cases a microscopical examination is lacking, and we are ignorant whether the mineral is actually composed of homogeneous crystals of one kind, or is an intergrowth of two or more crystalline phases. Moreover, as has become clear in the discussion of artificial crystals, a homogeneous crystalline mass may consist merely of a solid solution, the composition of which bears no necessary relation to that of any compound which may be formed by the component metals.

Nevertheless, chemical individuality is not to be denied to a mineral because no compound of similar composition is found by thermal analysis. Compounds occurring as minerals may have been produced in the wet way, by deposition from complex mixtures, under great pressures, or otherwise under such conditions as to yield a compound which has no range of stability within the limits of an ordinary investigation by means of thermal analysis, that is, by fusion of binary mixtures under atmospheric pressure. For example, the thermal analysis of the alloys of metals of high melting-point with arsenic is usually limited to alloys containing 50 per cent of arsenic or less, on account of the volatility of arsenic under atmospheric pressure, but metallic arsenides containing much larger proportions of arsenic are known to occur as minerals, and their chemical individuality is hardly doubtful, especially when their close analogy with the sulphides is taken into account. It must therefore be assumed that such minerals must have been produced under an increased pressure, or possibly by reactions occurring at a low temperature. Some of these native arsenides are stable when melted in a closed tube, but lose their excess of arsenic if heated under atmospheric pressure.

Of the metals enumerated above, those of high melting-point have a great tendency to form solid solutions, and such of their alloys as occur in nature are generally of that class. For example, the meteoric irons are composed of solid solutions of iron and nickel, whilst terrestrial alloys of those two metals are also known under the names of Awaruite and Josephinite. Gold occurs in the form of solid solutions with palladium, rhodium, and silver, the last alloy being known as electrum. Iridosmine contains iridium and osmium. Silver amalgams are occasionally found, but their composition does not correspond with the formula of any definite compound. A native alloy of gold and bismuth has been described as a compound, Au_2Bi, but it is improbable that any such compound exists, and the mineral may be a mere mixture.

The following arsenides and antimonides, which have been recorded as minerals, have formulæ corresponding with compounds already recognized by the method of thermal analysis :—

Dyscrasite, Ag_3Sb

Domeykite, Cu_3As

Niccolite, $NiAs$

Breithauphite, $NiSb$

Löllingite, $\Big\}FeAs_2$
Arsenoferrite,

The following minerals have received formulæ, but are probably solid solutions, more or less unsaturated :—

Allemontite, $SbAs_3$

(Variety of) Dyscrasite, Ag_6Sb

Whitneyite, Cu_9As

Algodonite, Cu_6As

Horsfordite, Cu_6Sb

3 *

Arsenargentite, represented as Ag_3As, and Chilenite, to which the formula Ag_6Bi has been given, may be mixtures.

The following minerals contain more arsenic than any compound of the corresponding series which has been studied by thermal analysis, but they probably represent definite compounds :—

Smaltite,	$\Big\}CoAs_2$
Safflorite,	
Chloanthite,	$\Big\}NiAs_2$
Rammelsbergite,	
Skutterodite,	$CoAs_3$
Sperrylite,	$PtAs_2$
Bismutosmaltite,	$Co(As, Bi)_3$

Leucopyrite, Fe_3As_4, is possibly a mixture of FeAs and $FeAs_2$.

The following well-defined tellurides occur in nature :—

Hessite,	Ag_2Te
Petzite,	$(Ag, Au)_2Te$
Tetradymite,	Bi_2Te_3
Coloradoite,	HgTe
Altaite,	PbTe

Of other tellurides, Stützite, Ag_4Te, may be a mixture of the compound Ag_2Te with silver, and Melonite, Ni_2Te_3, more probably contains the known compound NiTe. Joseïte is a variable bismuth telluride.

Sylvanite, $AgTe_2$, Calaverite, $AuTe_2$, and Krennerite (Ag, Au)Te_2, are recognized as definite minerals, but compounds corresponding with them have not been obtained by fusion, the only tellurides recognized in these series by thermal analysis being Ag_2Te, AgTe, and Au_2Te_3.

The complex minerals Kalgoorlite, $HgAu_2Ag_6Te_6$; Goldschmidtite, $AgAu_2Te_6$; and Wehrlite, a silver-bismuth telluride, may also be mentioned. No reference is here made to the very numerous minerals in which arsenic and tellurium are partly or wholly replaced by sulphur and selenium, distinctly non-metallic elements.[*]

It will be observed that all the compounds enumerated above are of a heteropolar character. Homopolar intermetallic compounds have not been observed to occur in nature.

[*] The names and assumed formulæ of the above minerals are taken from Dana's *System of Mineralogy.*

CHAPTER VI.

PHYSICAL PROPERTIES.

THE principal physical properties of alloys, including the density, hardness, conductivity for heat and electricity, thermoelectric power, magnetic susceptibility, etc., are profoundly modified by the formation of intermetallic compounds. In the absence of compounds, each such property is in general a continuous function of the composition in any given series of alloys. It may be an approximately linear function, as the specific volume of conglomerates and solid solutions, or it may pass through a maximum or minimum, as the hardness and electrical conductivity of solid solutions. Discontinuities may occur in cases of limited miscibility in the solid state, the curve which represents the variation of the property with composition then exhibiting an abrupt change of direction at the point at which a new phase makes its appearance. Failure to recognize this condition is responsible for the attribution of definite chemical formulæ to a large number of alloys, each of which is merely the limit of saturation of a solid solution. It is necessary to examine the relation of each individual property to the composition separately, and to determine in which cases the discontinuity of the function depends on the appearance of a new phase, and in which it presents itself at the concentration of the compound. This is done in the following sections.

A new field of metallography has been opened up by the detailed study, in a large number of concrete examples, of the dependence of the physical properties of alloys on their composition. The thermal and microscopical methods determine the nature of the heterogeneous equilibrium in metallic systems, that is, the limits of temperature and composition within which each phase is stable. These methods do not, however, give direct information as to homogeneous equilibria, that is, they do not indicate directly the nature of the molecules which compose any

37

single phase. This is, in fact, a question to which the doctrine of phases is incapable of furnishing an answer.

It is true that certain inferences may be drawn from the thermal observations. It has been assumed in the discussion in Chap. I that a maximum on the freezing-point curve is due to the formation of an intermetallic compound, the formula of which is given by the composition at which the maximum occurs. In the absence of any extended series of solid solutions there can hardly be any doubt as to the truth of such a conclusion, but ambiguity is possible when the branch of the curve on which the maximum occurs represents the freezing of solid solutions, as in the case of the alloys of lead and thallium, mentioned on p. 11. In such a case the thermal investigations leave the question open, except in so far as indications are afforded by an examination of ternary systems. The microscope does not furnish any decisive evidence on such a point. A microscopically homogeneous alloy may contain chemically different molecules without any obvious difference of structure.

Investigations, however, most of which have been carried out in recent years, have now proved that the determination of certain physical properties is capable of throwing light, not only on heterogeneous, but also on homogeneous equilibria. Such properties as the electrical conductivity and thermoelectric power, for example, are not only dependent on the number and relative proportion of the constituent phases, but also on the concentration of some one kind of molecule within a homogeneous phase. In such a case the property reaches a maximum or minimum for that composition of the phase at which the conce tration of the molecules in question is a maximum. Whilst, therefore, the density and some other properties vary continuously within the limits of miscibility in the solid state, the electrical conductivity and thermoelectric power only do so if the molecules composing the solutions increase or decrease continuously in number from the one limit to the other. Should the series include a definite compound, entering into solid solution with both its components, the property curve exhibits a marked discontinuity at the composition of the compound, that is, at the point at which the concentration of the compound is a maximum. The discontinuity is often such as to produce a sharp cusp on the curve.

Further, it is often possible, by an examination of the form of
the curve, to draw some conclusions as to the degree of dissociation of the compound within the phase, whether the latter be
liquid or solid. This branch of the study has not yet received
much attention.

The relationships dealt with in this section are very clearly
exhibited by the remarkable alloys of magnesium and cadmium,
to which resort has been had to establish several of the general
principles involved. These two metals form a single compound, CdMg (p. 16), which enters into solid solution with both
components and in all proportions. Its physical properties have
therefore been studied in some detail (pp. 43, 46). At lower
temperatures, the homogeneous solid solutions undergo resolution
into two constituents, and it is therefore possible to observe, in
a single series of alloys, the gradual change from a simple to a
complex system.

SPECIFIC VOLUME.

An alloy which consists of a simple conglomerate of its
component metals has a specific volume which bears a nearly
linear relation to its composition, and may therefore be calculated
from the specific volumes of the pure metals and their relative
proportions. The linear relation does not hold exactly, on
account of deviations from the condition of closest packing which
occur in an intimate mixture of two solid phases, such as a
eutectic alloy. These deviations are, however, small, and the
rule has been verified in a large number of instances, the slight
variation from a straight line, where noticeable, being of the
order expected.

The simple linear relation also subsists in the case of a continuous series of homogeneous solid solutions. On the other
hand, the specific volume-concentration curve deviates, sometimes to a very considerable extent, from the straight line when
a new solid phase, containing an intermetallic compound, makes
its appearance in the series. It most frequently happens that a
compound has a smaller specific volume than its components,
but the reverse condition is also observed. When only a
single compound occurs in a series, and solid solutions are
not formed to any large extent, the specific volume-concentration curve is made up of two straight lines, intersecting

at the composition of the compound, and thus enabling its formula to be inferred directly.   Where there is a limited mutual solubility in the solid state, a discontinuity, often very slight, and difficult to observe, occurs at the limit of saturation of each solid solution.   Thus, if an intermetallic compound forms solid solutions to a limited extent with each of its components, a condition of very frequent occurrence, the specific volume curve is made up of three straight lines.   The two points of intersection merely mark the compositions at which the new intermediate phase makes its appearance, and it is not possible to infer from them the composition of the compound.   It was formerly supposed that every intersection indicated the presence of a compound of that composition, and it has been a frequent practice to describe as definite compounds mixtures corresponding with breaks in a density or specific volume curve.

Where only a single intermetallic compound is present, and the curve is made up, as described above, of three intersecting straight lines, it is sometimes possible to obtain an indication of the probable formula of the supposed compound by producing the first and last of these lines, and observing the composition at which they then intersect.   If several compounds are formed, they may be separately indicated on the curve, or it may happen that only one of them deviates sufficiently from the mean to be recognizable, the others being formed from their components with comparatively little change of volume.

The principal data on this question are due to Maey,[71] who determined the specific volume of a number of series, and also recalculated the observations of others, who had mostly plotted the values of the density against the composition, a method which fails to represent the relations with sufficient simplicity.   The best of these older determinations are due to Matthiessen.[72]

In the following cases the compound is formed with contraction, and the curve is composed of two straight lines :—

System.	Compound.	Per Cent Contraction.
Ag – Sb	Ag_3Sb	5
Ag – Sn	Ag_3Sn	5
Fe – Sb	Fe_3Sb_2	16

In the following series, the contraction due to the compound is distinctly marked, but the two branches of the curve are not straight lines :—

Cu – Sn	Cu_3Sn	10
Cu – Zn	Cu_2Zn_3	4

A small contraction was observed, leading Maey to propose a formula for a compound, in the series gold-bismuth and bismuth-lead, as well as in the cadmium amalgams, which are now known to be composed of two solid solutions. A curve composed of three branches, from which the formula of a compound may be obtained by prolonging the two end branches until they intersect, is given by the alloys of copper and antimony :—

$$Cu – Sb \qquad Cu_3Sb \qquad \text{Contraction, } 11\cdot5 \text{ per cent.}$$

The case becomes more complicated when the form of the curve is such as to suggest the presence of more than one compound, and it is only possible to infer the formulæ of the corresponding compounds in the absence of any extended miscibility in the solid state. The following breaks have been observed :—

Hg – K	Hg_9K ?	Hg_3K	Hg_2K	HgK
Hg – Li	Hg_5Li	Hg_3Li	HgLi	$HgLi_3$
Hg – Na	Hg_5Na ?	Hg_2Na	HgNa	$HgNa_3$
Ag – Cd	?	Ag_2Cd_3		

The maximum contraction in each series is due to the compound in italics.

Instances of compounds having a specific volume greater than that of their components are less frequent; but well-defined curves, composed of two intersecting straight lines, have been obtained for :—

System.	Compound.	Per Cent. Expansion.
Cd – Sb	Cd_3Sb_2	10
Sb – Zn	Sb_2Zn_3	7
Al – Sb [73]	AlSb	24

A good example of a system containing two compounds, each of which is formed with expansion, is found in the alloys of cadmium and arsenic :—[74]

	Per Cent Expansion.
Cd_3As_2	$19\cdot5$
$CdAs_2$	$14\cdot4$

Solid solutions are not formed to any appreciable extent in this case, so that the curve of specific volumes (or, better, of atomic volumes) consists of three straight lines, intersecting at compositions corresponding with the two compounds. Cd_3As_2 appears as a maximum on the curve, $CdAs_2$ only as an abrupt change of direction.   The constitution of the alloys has been definitely established by thermal and microscopical analysis.

It is worthy of remark that the compounds which have been proved to be formed from their component metals with increase of volume are either antimonides or arsenides.

As an example of the somewhat hazardous attempts to investigate the constitution of complex series by the method of specific volumes, mention may be made of a study of metallic silicides, in which the discontinuities observed were more numerous than would be expected from the thermal data.[75]

HARDNESS.

The hardness in a series of conglomerates is approximately proportional to the composition, the slight deviations usually observed being due to the fact that the constituents of a fine-grained eutectic support one another, so that the hardness as determined by most ordinary methods is slightly increased.   On the other hand, the hardness in a series of solid solutions is much greater than that calculated by the rule of mixtures, and the hardness-composition curve commonly passes through a maximum. The hardness of an alloy containing equal parts of silver and gold, for example, is nearly twice that of either of the component metals.[76]

An intermetallic compound is usually harder than its components, although this cannot be stated as an invariable rule. The difference is often very great, the extreme hardness of such a compound as Cu_3Sn, for example, being very remarkable.

The methods adopted for the determination of hardness, although differing widely in character and theoretical significance, have comparatively little influence on the form of the curve connecting hardness and composition, provided that all the alloys are examined in the crystalline condition without being strained by the application of mechanical work.   Four such methods are in general use in the investigation of this property, namely, the sclerometric method, depending on the production of a scratch of measurable breadth ; the indentation method, in which the extent of the yielding under the pressure of a loaded ball or cone is measured; the elastic reaction method, in which the coefficient of restitution is determined when a hard object falls on to the specimen from a height; and the plastic flow method, in which the material is forced through an orifice under the application of very high pressures. Unlike as these four methods of testing appear to be, they yield remarkably concordant results when applied to normal crystalline alloys. The third is the least valuable for this purpose, whilst the first and second are very generally applicable, and the fourth, which is impracticable except in the case of relatively soft materials, has proved of value in the theoretical comparison of hardness with other properties.

A series of alloys containing a single compound, and free from solid solutions of appreciable concentration, has a hardness curve composed of two straight lines intersecting at the composition of the compound. A compound which forms solid solutions in all proportions with both components has its hardness increased by the addition of either metal, and the curve thus has two maxima, the composition of the compound being indicated by a cusp directed downwards (Fig. 6). This case presents itself in the alloys of magnesium and cadmium at temperatures above their transformation-point (Fig. 7). However, the conditions in this series are somewhat less simple, as the compound CdMg, which in its β state is miscible in all proportions with cadmium and magnesium, has only a limited miscibility in the a-condition, so that determinations of the hardness of the slowly cooled alloys yield a rather more complex curve, with minor discontinuities at the limits of saturation of the solid solutions.[77] The cusp directed downwards indicates unmistakably the position of the compound, in a case in which the thermal analysis leaves the existence of a compound unproven, although probable, whilst the microscopic structure of the entire series is, apart from transformations at a lower temperature, completely homogeneous.

A more complex series is seen in the alloys of magnesium and silver (Fig. 3). The equilibrium diagram in the upper part of the figure[33] shows that two compounds are formed, of which the first, $AgMg_3$, decomposes below its melting-point and does not enter into solid solution to any appreciable extent, whilst the second, AgMg, appears as a maximum on the freezing-point curve, and forms solid solutions with both silver and magnesium. The hardness curve, shown immediately below, indicates that $AgMg_3$ has a hardness, measured by Brinell's ball method, more than six times as great as that of either magnesium or silver. Silver is hardened in a very marked

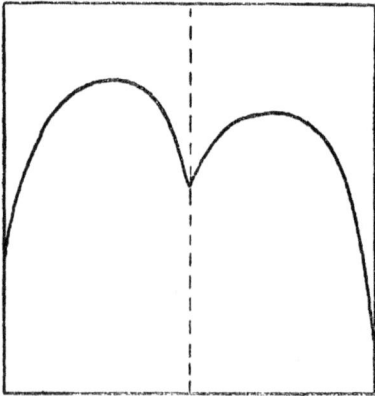

Fig. 6.—Hardness curve.

degree by the entry of magnesium into solid solution with it, thus causing the second maximum in the curve. The compound, AgMg, appearing as it does in the midst of a series of solid solutions, is indicated by a cusp directed downwards. It is nevertheless nearly three times as hard as its components.[78]

These two curves are typical of those obtainable from alloys containing intermetallic compounds, and it is not

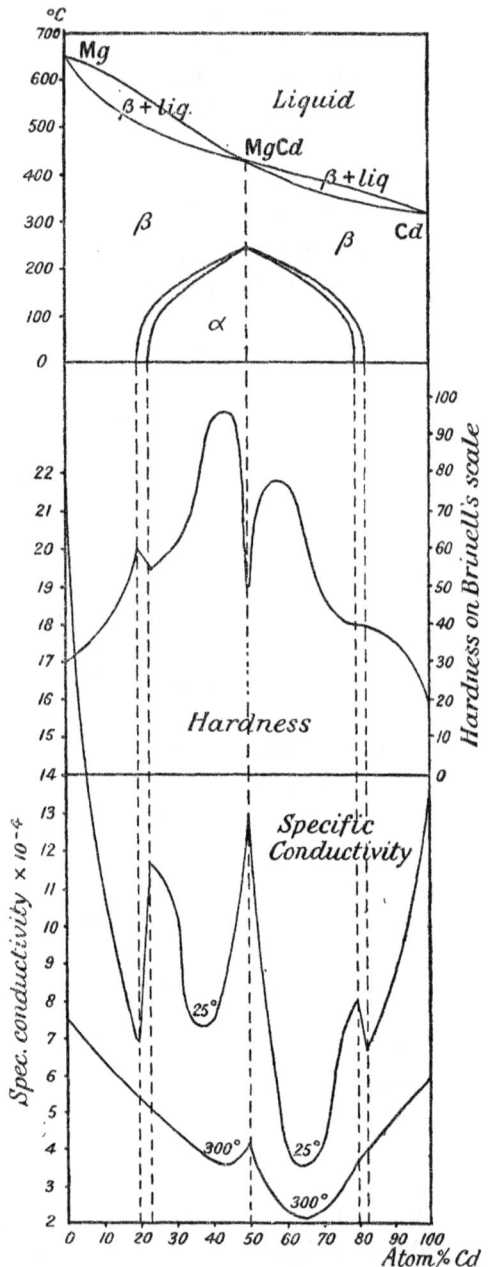

Fig. 7.

necessary to cite further examples in detail. The technically important alloys of copper with zinc, tin, and aluminium, on account of their highly complex constitution, yield less satisfactory curves, but a few conclusions may nevertheless be drawn from them. The alloys of copper and zinc have a maximum hardness at the composition of the compound Cu_2Zn_3, which is ten times as hard as copper, using either the Brinell ball test or the elastic reaction method.[79] The compound Cu_3Sn is determined by the scleroscopic method to be twelve times as hard as copper.[80] The compound Cu_3Al is also much harder than its components, being about seven times as hard as copper.[61]

ELECTRICAL CONDUCTIVITY.

The general laws connecting the electrical conductivity with the constitution of alloys have been well established by the labours of a number of investigators, beginning with Matthiessen.[1] Alloys which consist of simple conglomerates of the component metals have a conductivity which bears an almost linear relation to the composition by volume, the small deviations which are frequently observed being of the order of perturbations due to the mechanical arrangement of the constituent phases. When a continuous series of solid solutions is formed, the conductivity curve passes through a minimum, the fall being extremely steep in the neighbourhood of the pure metals. When the concentration of the solid solutions is limited, an abrupt change of direction is observed at the limits of saturation.

The presence of intermetallic compounds profoundly modifies the form of the conductivity curve. The principal relations have been stated by Guertler,[81] and by Kurnakoff and Schemtchuschny.[76]

Whilst the composition by volume is undoubtedly the proper basis of a quantitative comparison of the conductivity of conglomerates or of solid solutions, it is preferable to use the atomic percentages when the object is to determine whether intermetallic compounds are present or not in a given series of alloys. For the systematic study of a series, it is desirable to construct a diagram in the first instance having atomic percentages as abscissæ, and then, in the case of any singularities pointing to the presence of intermetallic compounds being observed, to divide the system into as many binary systems as are thus indicated, and to construct a corresponding number of partial diagrams, in which percentages by volume are employed as abscissæ.

The further question, whether the specific resistance or specific conductivity should be adopted for the ordinates of the diagram, can only be answered by saying that both may be employed. As a rule, any discontinuities are more clearly marked on the curve of conductivity than on that of resistance, but the contrary case sometimes presents itself when the compound occurs near to a minimum in the conductivity.

An intermetallic compound, if well defined, may be regarded conveniently as a separate component, forming alloys with the pure metals or with other compounds in the same series. The conductivity of a compound is always less than that of the better conducting of the two metals of which it is composed. As the conductivity of a pure metal is always lowered by alloying with a second metal with which it forms solid solutions, the same may be expected to hold good of an intermetallic compound. Should the compound be capable of entering into solid solution with both

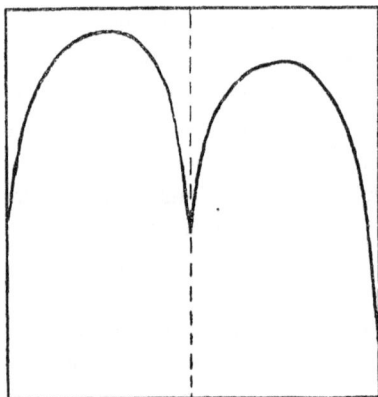

FIG. 8.—Conductivity curve.

of its components in all proportions, the conductivity curve should have the form shown in Fig. 8, the compound being indicated by a cusp, since its conductivity is necessarily lowered by an excess of either component. This case is met with in the alloys of magnesium and cadmium at temperatures above their transformation point (Fig. 7). The curve showing the conductivity of the alloys at 300° refers to the homogeneous series of β-solid solutions. The compound MgCd occurs in the middle of this series. Its conductivity is not much more than half that of magnesium at the same temperature, and is depressed by alloying with either of the components. The conductivity diagram at 300° thus consists of two approximately U-shaped curves, meeting in a sharp cusp directed upwards, at the composition MgCd.[77]

At a lower temperature, such as 25°, the β-solid solutions of medium concentration have undergone conversion into the α-modification, which is restricted within certain limits of concentration. There are now two ranges of stable β-solutions, adjoining the

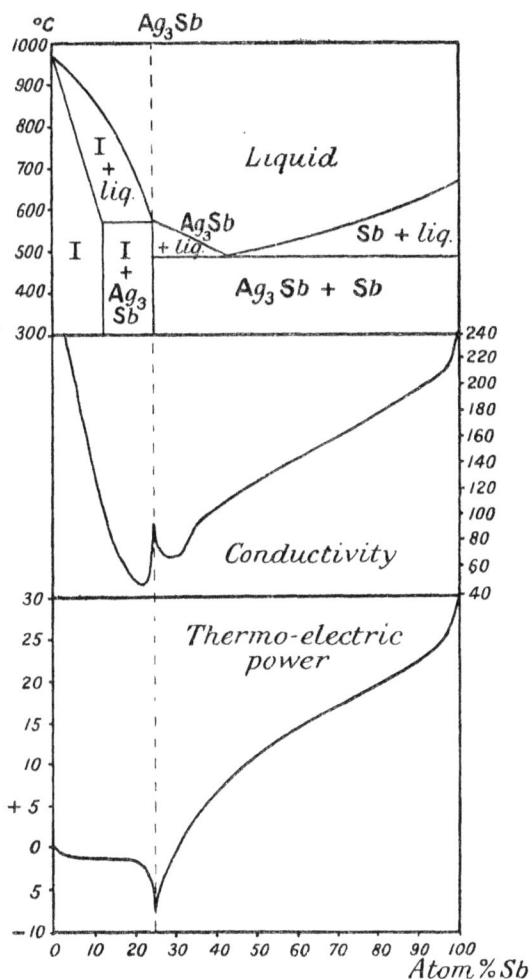

FIG. 9.

magnesium and cadmium ends of the series respectively, a middle range of stable α-alloys, and two intermediate regions in which α and β are present in equilibrium with one another. It is now possible to predict the form of the conductivity-composition curve for 25°. The *a*-region corresponds with two U-shaped curves uniting in a cusp directed upwards, the two *a* + *β*-regions with straight lines, whilst the terminal *β* solid solutions are indicated by conductivity curves falling steeply from the values for the

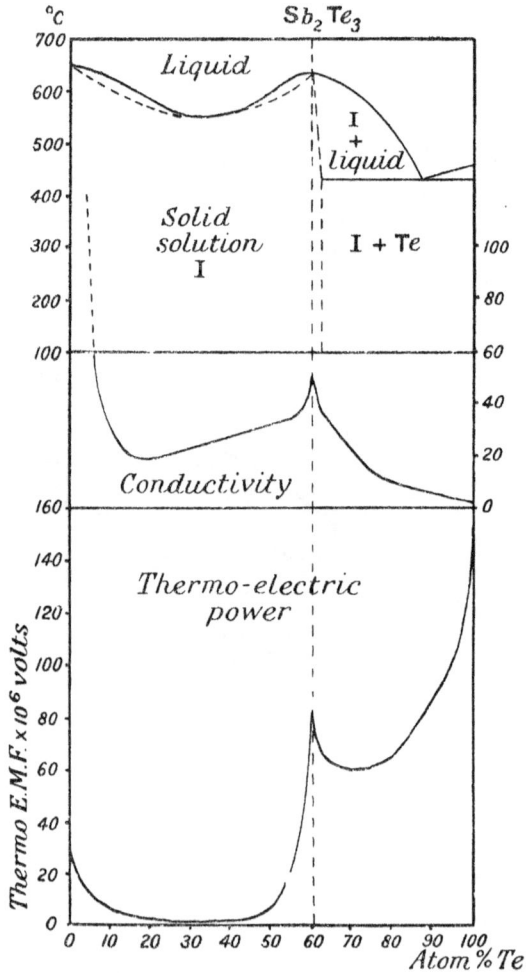

FIG. 10.

pure metals. This is the actual form of the curve determined experimentally. At temperatures intermediate between 25° and 300°, the *a*-region diminishes in extent with rising temperature. The form of the transformation curves has in fact been determined more accurately by electrical than by thermal means.

The simplest possible system in which an intermetallic compound occurs is one in which solid solutions are not formed to any appreciable extent. The series may then be split up into two simple binary systems, each composed only of conglomerates.

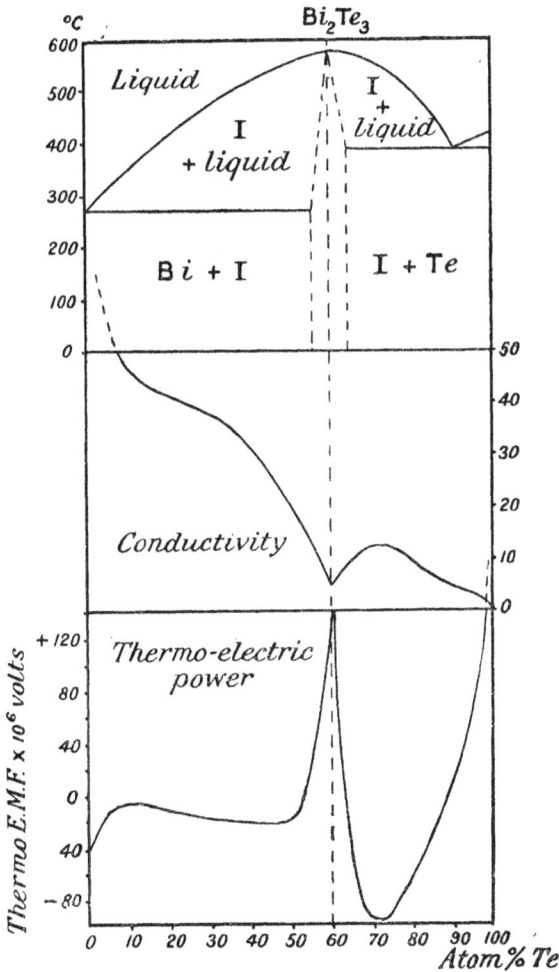

FIG. 11.

The conductivity is represented by two straight lines, intersecting at the composition of the compound. The closest approach to this condition has been found in the alloys of magnesium and bismuth. The compound Mg_3Bi_2 has a conductivity very near

4

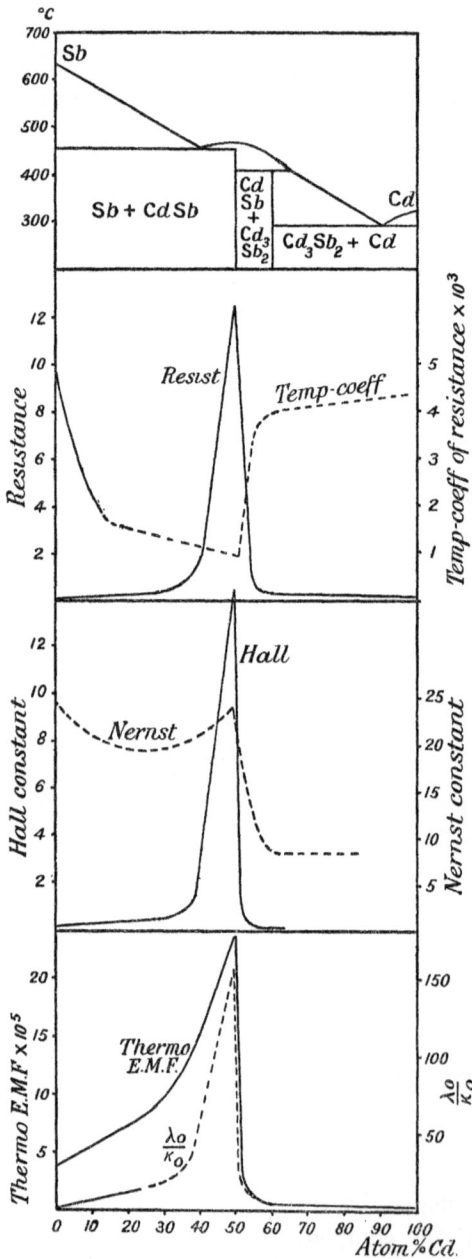

Fig. 12.

that of bismuth, and the two series Mg-Mg_3Bi_2 and Mg_3Bi_2-Bi are simple conglomerates. On the other hand, the conductivity curves of the series magnesium-lead, magnesium-tin, and magnesium-zinc, all indicate the formation of solid solutions to an appreciable extent at the magnesium end of the series.[82]

Other systems are represented in Figs. 9, 10, and 11. The compounds Ag_3Sb and Sb_2Te_3 are clearly indicated by cusps directed upwards. On the other hand, the cusp corresponding with Bi_2Te_3 is directed downwards. The reason for the maximum on the tellurium side is not evident, but it appears to point to imperfect equilibrium in the alloys used for the measurements. It is remarkable that the curve of thermoelectric power, shown in the lower part of the diagram, does not exhibit any irregularity at this point, but has an exceptionally sharp cusp, with a smooth U-shaped curve between the compound and the tellurium end of the series.

The alloys of antimony and cadmium (Fig. 12) have been plotted with the resistance instead of the conductivity, as the compound SbCd is more clearly marked in this way. An exactly similar curve is given by the alloys of antimony and zinc, with the cusp at the composition of the compound SbZn.[83]

The only rule which has been enunciated with respect to the relative conductivities of intermetallic compounds is, that the conductivity is lower, the less electro-positive one of the components of the compound is. Thus antimonides and arsenides have a low conductivity, and in fact approach non-metallic substances in their electrical properties.

The temperature-coefficient of the resistance or of the conductivity is as characteristic a function of the constitution of an alloy as the conductivity itself. Like the conductivity, it varies in a linear manner in a series of conglomerates, and along a curve passing through a minimum in a series of solid solutions. Alloys having a temperature coefficient near to zero, such as are used for the construction of resistance coils, are always solid solutions, near to such a minimum.

The temperature-coefficient of the resistance is approximately the same for all pure metals, being always positive, and usually from 0·003-0·004 at about atmospheric temperatures. The co-efficient is not a constant, but decreases with rising temperature. A complex formula is necessary to express the relation of resistance to temperature over any extended range.

Homopolar intermetallic compounds are found to have temperature coefficients which are approximately the same as those of pure metals. They are thus clearly distinguished from those alloys which consist merely of solid solutions of their components. This is sufficiently indicated by the following table, in which an alloy consisting of a solid solution is selected from each series for comparison, being placed in the right-hand part of the table [82] :—

Series.	Compd.	Temp.-coeff. $a \cdot 10^5$.	Solid soln. Atomic per cent.	Temp.-coeff. $a \cdot 10^5$.	Reference.
Mg-Pb	Mg_2Pb	250	95 Mg, 5Pb	65	
Mg-Sn	Mg_2Sn	445	96 Mg, 4Sn	110	
Mg-Cu	Mg_2Cu	365	8 Mg, 92 Cu	120	82
...	$MgCu_2$	316	...	...	
Mg-Bi	Mg_3Bi_2	370	...	...	
Mg-Zn	$MgZn_2$	290	35 Mg, 65 Zn	97	
Ag-Mg	Ag_3Mg	309	70 Ag, 30 Mg	161	78
...	AgMg	310	22·5 Ag, 77·5 Mg	159	
Cd-Mg	CdMg	417	...	...	77
Sb-Sn	SbSn	343	...	...	84
...	Sb_2Sn_3	361	...	...	
Cu-Sn	Cu_3Sn	350	90 Cu, 10 Sn	43	85
...	CuSn	300	...	...	
Cu-Zn	CuZn	360	70 Cu, 30 Zn	160	
...	$CuZn_2$	408	41 Cu, 59 Zn	200	86
...	$CuZn_6$	355	21 Cu, 79 Zn	240	
Cu-As	Cu_5As	250	94 Cu, 6As	32	87
Al-Mg	Al_3Mg_4	405	37 Al, 63 Mg	30	88

The very low values for the coefficient of the resistance in the case of solid solutions are in striking contrast with the uniformly high values for intermetallic compounds.

The curves for a complex system, the alloys of copper and zinc, are given in Fig. 13.[86]   The resistance curve has two cusps, at compositions corresponding with the compounds Cu_2Zn_3 and $CuZn_6$. The maximum in the first part of the curve is that due to the a-solid solution. The conductivity curve has two cusps, one of which is placed exactly at the composition CuZn, whilst the other is more nearly at $CuZn_5$. A third discontinuity perhaps represents $CuZn_9$. On the temperature-coefficient curve the four upward cusps correspond with CuZn, $CuZn_2$, $CuZn_6$, and $CuZn_9$ respectively. Pushin therefore regards the compounds CuZn, $CuZn_2$, and $CuZn_6$ as certain, and $CuZn_9$ as possible. This involves the rejection of Cu_2Zn_3, a compound for which there is much other evidence, but which Pushin regards merely as the limiting concentration of the γ-phase, containing $CuZn_2$. There does not seem to be any good reason for this. The point $CuZn_2$ is not marked on the conductivity curve, although it appears so conspicuously on the temperature-coefficient curve. The low coefficient of Cu_2Zn_3 may be accounted for by supposing it to approach the antimonides and other heteropolar compounds more closely than the other members of the series, a view which receives support from other facts.

An interesting feature of the curves is the prominence of the compound CuZn, the central constituent of the β-phase, the existence of which has been sometimes questioned.

The electrical behaviour of alloys at low temperatures is best

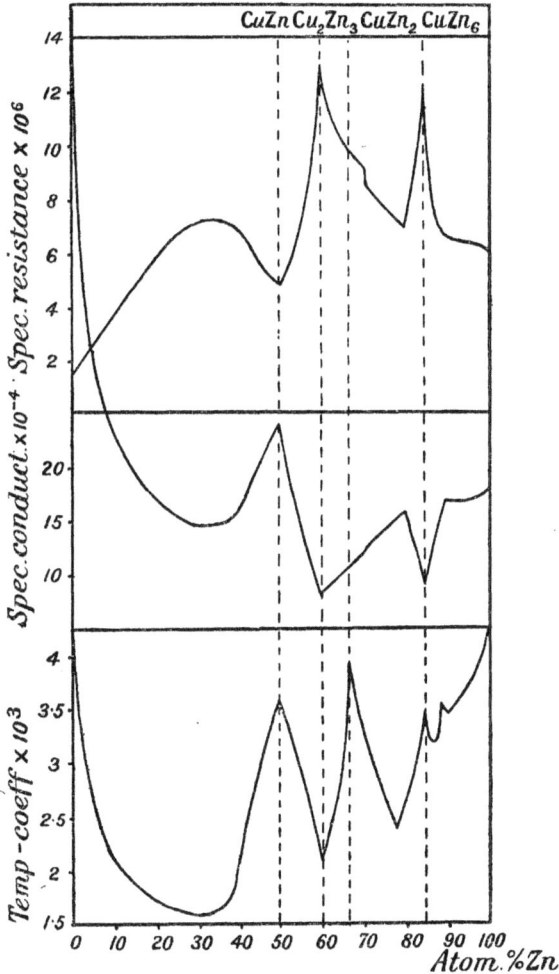

Fig. 13.

examined by considering the resistance instead of the conductivity. Much recent work [89], [90] goes to confirm the view that the electrical resistance of pure metals becomes zero at or somewhat above the absolute zero of temperature. Solid solutions, on the other hand, exhibit a resistance which indeed diminishes with falling temperature, but reaches a finite constant value instead of vanishing at very low temperatures. The "solution resistance," or additional resistance above that calculated by the rule of mixtures, due to the formation of a solid solution is in fact independent of the temperature.

Semi-conductors (non-metals, metallic oxides and sulphides, etc.) behave somewhat differently. Their resistance passes through a minimum at a definite temperature, below which the temperature-coefficient of the resistance is negative, whilst above it it is positive. Thus magnetite, Fe_3O_4, has an inversion temperature of $150°$. It is likely that the more heteropolar intermetallic compounds have similar properties. Should this be confirmed when a sufficient number of such compounds have been examined over a wide range of temperature, it would be an interesting addition to the evidence in favour of the view that such compounds approach the non-metals in their properties.

THERMAL CONDUCTIVITY.

The thermal conductivity of alloys follows similar laws to the electrical conductivity, but is much less conveniently measured. The thermal conductivity of a few intermetallic compounds has been determined. In the antimony-cadmium series [91] the conducting power of the compound SbCd for heat is little more than that of window glass, and about one-fifth of that for quartz or felspar. Moreover, the temperature-coefficient of the thermal conductivity $\frac{\lambda_{-190°}}{\lambda_{0°}}$ has a value of $1\cdot2$-$1\cdot5$ for metals, whilst for SbCd it has the value $2\cdot8$, being about 3 for non-conducting crystals.

The ratio of thermal to electrical conductivity is also an important quantity. The ratio $\frac{\lambda}{\kappa}$ has almost the same value for pure metals at a given temperature, and increases proportionately to the absolute temperature.[92] The ratio $\left(\frac{\lambda}{\kappa}\right)_{100°} : \left(\frac{\lambda}{\kappa}\right)_{0°}$ has a value for pure metals approaching $1\cdot367$.[93]

$\frac{\lambda}{\kappa}$ is regularly higher for solid solutions than for pure metals, and may be much higher for intermetallic compounds. The compound SbCd (Fig. 12) which approaches the non-metals in its properties, has $\frac{\lambda_0{}^\circ}{\kappa_0{}^\circ} \cdot 10^6 = 158$, a value comparable with that for carbon, 180.[91]

THERMO-ELECTRIC POWER.

Although many determinations of the thermo-electric power of alloys have been made since the original observations of Seebeck in 1826, and although some of the results obtained have played a prominent part in discussions relative to the conduction of electricity in metallic solid solutions, little attention has been devoted to the effect of intermetallic compounds on this property. A communication by Haken,[94] however, contains results which show that the singularity in a curve due to the presence of a compound is sometimes exceptionally well marked in the case of the thermo-electric power. The curves here reproduced are taken from Haken's results.*

In general, the thermo-electromotive force of a series of binary alloys varies with the composition in the same manner as the electrical conductivity. The variation is linear in conglomerates of the pure metals, exhibits sudden changes of direction if solid solutions are formed to a limited extent, and has the typical U-shape if the two metals form an unbroken series of solid solutions.[95] These rules have been verified in a number of instances.

Haken's investigation relates mainly to series in which one or more intermetallic compounds are present, and the form of the curves indicates that this property is even more sensitive than the electrical conductivity in detecting miscibility of one or other component with the solid binary compound. As an example, the system bismuth-tellurium (Fig. 11) may be considered. The thermal diagram, shown in the upper part of the figure, had been obtained independently by two investigators,[96, 97] neither of whom observed the formation of solid solutions, although their measurements were not sufficiently exact to prove the complete absence of such a condition. The maximum on the freezing-point curve points decisively to the existence of a stable compound, Bi_2Te_3. The same compound is indicated in a most striking manner on

---

* The curves have been re-plotted, using atomic percentages as abscissæ in place of percentages by weight.

the curve shown in the lowest compartment of the figure.* The interpretation of such a diagram is not difficult. The steep rise of the curve at the bismuth end of the series shows that solid solutions in bismuth exist to a small limiting concentration. The sharp cusp at the composition of the compound, the two branches of the curve then falling away with remarkable steepness, indicates that the compound Bi_2Te_3 is capable of retaining a small excess of bismuth or tellurium in solid solution. From the limiting concentration of this solution on each side the thermo-electromotive force varies in an almost linear manner with the composition, although the errors of experiment are somewhat too great to allow of the determination of the exact points of intersection of the respective branches.

It appears that the electrical conductivity in this instance, as represented in the middle compartment of the figure, although showing quite clearly the position of the compound, is less sensitive as an indication of the formation of solid solutions.

The next example, that of the alloys of antimony and tellurium (Fig. 10), is of interest as confirming a somewhat unusual form of equilibrium diagram arrived at as the result of thermal analysis. The freezing-point curve [98] has a maximum corresponding with the compound Sb_2Te_3, whilst on the antimony side the curve is continuous, passing through a minimum, showing that antimony and its telluride are completely isomorphous. On the tellurium side the freezing-point curve has the form usual in a eutectiferous series. The electrical properties of the alloys are in complete accordance with these conclusions. The curve of thermo-electromotive force has a sharp cusp at the composition Sb_2Te_3, and a smooth U-shaped branch between that compound and antimony, whilst the short descending branch on the tellurium side indicates that solid solutions with that element are only formed to a very limited extent. The extreme sharpness of the cusp is noticeable, a fall of 0·4 per cent in the proportion of tellurium causing a diminution of the *E.M.F.* from 83 to 50.

In the third example selected, that of the alloys of silver and antimony (Fig. 9) the conditions are so far different that the only compound present in the series, Ag_3Sb, is broken up below its

---

* The metal used for comparison throughout these experiments was copper. One junction was maintained at 19° and the other at 0°. The *E.M.F.* is calculated for 1° difference of temperature.

melting-point, forming a solid solution rich in silver, and a liquid.[99] The discontinuity in the freezing-point curve occurs exactly at the composition of the compound, an exceptional case which also presents itself in the closely allied gold-antimony series.[100] The thermo-electric curve has a very distinct cusp, in this instance directed downwards, the *E.M.F.*, measured against copper, in the neighbourhood of the compound having the opposite sign to that determined in the other series. The form of this curve and of the curve of conductivity indicates very distinctly that the compound forms solid solutions with both components within certain limits, a fact overlooked in the thermal and microscopical study of these alloys.

The other series investigated in the same manner by Haken include the following: Tin-tellurium, with a small cusp at the composition SnTe, suggesting that this compound is largely dissociated and that solid solutions are only formed to a minute extent, conclusions which are entirely in accordance with the flat form of the maximum on the freezing-point curve, and with the thermal and microscopical observations generally.[101, 102] Lastly, the alloys of bismuth and antimony gave a curve which furnished no indication of the presence of any compound, the irregularities observed being evidently due to the difficulty of obtaining homogeneity in the readily fusible solid solutions of which the whole series consists.[103] The existence of the compound PbTe [104] was confirmed, but the physical properties of the alloys of this series rendered them unsuitable for a complete study.

Another system which has been investigated in detail is that of cadmium and antimony.[83] The equilibrium diagram of this series is peculiar, in that two compounds are formed, but the appearance of one of them, CdSb, is readily suppressed by under-cooling, as described previously (p. 21). This compound is nevertheless marked by a very sharp cusp on the curve of the thermo-electromotive force (Fig. 12), indicating that the alloys used for the determinations were in a condition of equilibrium.

ELECTROLYTIC POTENTIAL.

The electrolytic potential of an alloy, like any other of its physical properties, is a function of its composition, but the form of the function is not always readily determined. The difficulties of determination are partly theoretical and partly experimental. The theory of the equilibrium between an alloy and an electrolyte has only been established for the simplest cases, those of solid solutions and of simple conglomerates, whilst the manner in which an intermetallic compound, or a solid solution containing such a compound, dissolves (in other words, the nature of the ions which it sends out into the solution) remains almost entirely unknown. The experimental difficulties consist in finding an electrolyte with which the alloy can be in equilibrium, and in overcoming the disturbing effect of superficial changes in the composition of the alloy.

Certain conclusions as to the conditions of equilibrium have been reached by Reinders,[105] by means of an application of the phase rule. Reinders' formulæ have been verified in the case of the cadmium amalgams, which consist of two series of solid solutions, separated by a gap. Determinations of electrolytic potential have, in fact, been employed with success to fix the limits of the gap at different temperatures.[106] The limits of saturation of the two solid solutions are clearly marked by discontinuities in the curve of electrolytic potential.

In the case of a compound, the assumption made by Reinders, for purposes of calculation, is that the ions sent out preserve the same ratio of the two metals as in the compound. This assumption is almost certainly untrue for many alloys, the compound being resolved into its components at the surface of contact with the electrolyte, so that only one metal passes into solution. Even with this assumption, it is difficult to determine the composition of an electrolyte which shall be in true equilibrium with the alloy, and in most actual investigations a more or less arbitrary composition of the electrolyte has been adopted.

The earliest investigations having any claim to exactness are those of Laurie, who neglected one important factor, the concentration of the electrolyte, but adopted precautions to reduce the effect of polarization to a minimum. Thus, the copper-tin alloys were studied by constructing a cell with a porous partition, the alloy being immersed in a solution of stannous chloride in one compartment, whilst the other electrode was a rod of copper immersed in a solution of copper sulphate. In this way a discontinuous *E.M.F.* curve was obtained, with a well-marked break at the composition Cu_3Sn.[107] In a similar manner, the alloys of gold and tin were found to exhibit a discontinuity at the composition AuSn.[108]

The later work of Herschkowitsch [109] is more extensive. In this investigation, the electrolyte was in each case a normal solution of the more positive metal, whilst the electrode for comparison was a rod of the less positive metal, an arrangement which is open to objection, and the results obtained were in some cases ambiguous.

Pushin [110] adopted the plan of employing as an electrolyte, wherever possible, an acid or alkali which forms a sparingly soluble salt with the more positive metal, the ionic concentration of which in the solution is thus kept as low as possible.

The simplest case is that in which the two metals unite to form a single compound, which does not enter into solid solution with either of its components. The potential of the alloys is then that of the more positive metal, as long as any of the latter is present as a distinct phase. The disappearance of this phase coincides with the composition of the compound, and at this point the potential falls abruptly to a lower value. If two or more compounds are formed, a corresponding number of abrupt falls of potential may be observed. On the other hand, if solid solutions are formed, the potential varies continuously within the limits of any solid solution, and only varies abruptly at the point at which a phase disappears. It may be impossible in such a case to infer the composition of the compound with certainty.

A simple example is presented by the thallium amalgams,[111] shown in the upper part of Fig. 14. The potential in this example is compared with that of a mercury electrode, and the *E.M.F.* of the combination rises gradually from zero to 33·3 atomic per cent of thallium, beyond which it remains constant, indicating that the compound Hg_2Te forms simple conglomerates with thallium, but enters into solid solution with mercury.

A few of Pushin's curves are redrawn in the lower part of Fig. 14, from which it will be seen that certain discontinuities are very distinctly marked. The points are not always sufficiently close together to allow of the determination of the true form of the curve, and in some instances changes have probably been represented as abrupt which on a more careful examination would prove to be gradual. As a means of detecting intermetallic compounds, the method may give results which err in

either of two directions. On the one hand, the occurrence of
solid solutions may give rise to discontinuities indicating more
compounds than actually exist; on the other, a compound may
have an electrolytic potential which differs so little from that of
one of the components that the change in direction of the curve
is imperceptible. The former case is exemplified in the list
below by the alloys of silver and zinc, and the latter by those
of silver and tellurium.

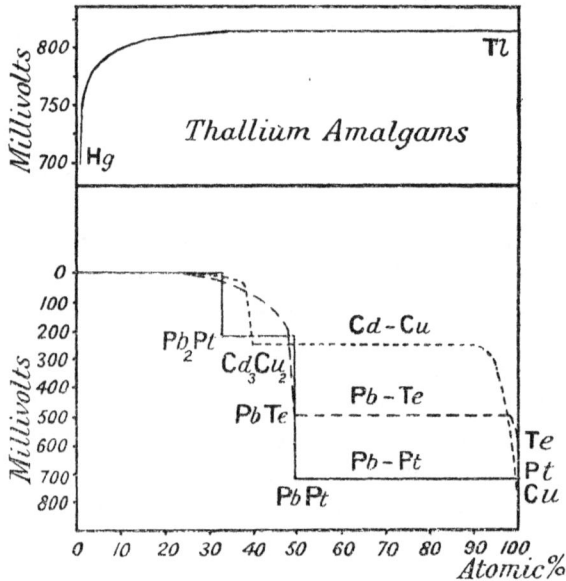

Fig. 14.

The following compounds, the existence of which has been
sufficiently established by other methods, are indicated on
Pushin's curves of electrolytic potential, although in some cases
the curves have been wrongly interpreted. A correction has
now been made by comparison with the thermal diagram :—

Ag_3Sb—Slightly indicated. The distinct break at the composition
Ag_2Sb merely marks the limit of saturation of the solid
solution.

Ag_3Sn—Very distinct. Limit of solution also indicated.

Ag_2Te—Very distinct. The potential of this compound is so
near to that of tellurium that the second compound,
AgTe, is not indicated.

Ag_2Zn_3—Wrongly interpreted as $AgZn_2$.

Ag_2Zn_5—Probably indicated on the curve. The remaining breaks only mark the limits of solid solutions.

Al-Cu system—A single break occurs near 50 atomic per cent, attributed to a compound AlCu, but really marking the limit of Al_2Cu.

AsSn—Observed only in contact with acid electrolyte.

As_2Sn_3—Observed in acid and alkaline electrolytes.

AuSn—Very distinct.

$AuSn_2$—A small but distinct break. The compound $AuSn_4$ is not indicated.

AuZn—Very great change in potential.

$AuZn_2$—Indistinctly marked. A further compound, $AuZn_6$ or $AuZn_8$, is faintly indicated.

Cd_3Cu_2—Erroneously given as Cd_2Cu, this being near the limit of saturation of solid solution. The compound $CdCu_2$ is entirely without influence on the curve.

Cu_3Sn—Very distinct. The only other compound found is Cu_2Sn, and this is almost certainly erroneous.

Cu_2Te—Very distinct. A compound CuTe is also inferred, but finds no justification on the curve.

Cu_2Zn_3—This compound is probably the cause of the very distinct break attributed to $CuZn_2$. The other compounds inferred mark the limits of solid solutions.

Pb_2Pd—The other existing compounds are not indicated.

Pb_2Pt—Distinct.

PbPt—Distinct. The potential of this compound is close to that of platinum, and the third compound is not indicated.

PbTe—Distinct. A simple case, with one break.

SbNi—Distinct.

Sb_2Ni_5—Recorded as $SbNi_3$. Equilibrium in these alloys is not readily attained.

SbSn—Distinct.

Sb_2Sn_3—Only slightly marked.

SbZn—Very distinct.

Sb_2Zn_3—Distinctly indicated.

SnTe—Very distinct. A clear and simple example.

HEAT OF FORMATION.

The direct calorimetric determination of the heat of formation of intermetallic compounds has rarely been attempted. A few calorimetric measurements were made by Person,[112] but none of the series examined by him contained compounds. Berthelot[113] showed that the heat-change on dissolving silver in mercury did not indicate the production of a compound having any appreciable heat of formation.

Indirect methods have been frequently applied. Thus, Berthelot[114] prepared amalgams of sodium and potassium, and measured the heat evolved when they were decomposed by means of a dilute acid. Plotting his observations, and allowing for the different atomic weights employed by him, it is found that a maximum heat of formation occurs at the compositions Na_6Hg and $K_{12}Hg$ respectively. This is not in accordance with the results of subsequent thermal analysis, and the determinations have not been repeated with modern precautions.

The heat of formation may be determined indirectly by measuring the heat developed when alloys containing the compound in question are dissolved in some suitable reagent, and comparing it with that which would be obtained by dissolving the component metals separately under the same conditions. This method was applied to the alloys of zinc and copper, dilute nitric acid being used as the solvent.[115] This procedure was shown to be illegitimate,[116] as the chemical reactions obtained vary with the composition of the alloy, different nitrous gases being obtained as the proportion of zinc varies. The same and other alloys were also examined by Herschkowitsch,[109] who used a solution of bromine in potassium bromide as solvent, thus avoiding the difficulty just mentioned, but the number of his observations is insufficient to establish any definite conclusions. The same solvent was employed in a study of the alloys of copper and aluminium,[117] the inference being drawn that the maximum heat of formation corresponded with a compound Cu_2Al, for which there is no other evidence. The experimental results are, however, quite consistent with the assumption that the principal heat of formation is that of the known stable compound Cu_3Al. Very irregular results were obtained by dissolving alloys of aluminium and zinc in dilute hydrochloric acid.

The best calorimetric determinations of this kind refer to the alloys of copper and zinc.[118] The solvents used were solutions of ferric ammonium chloride and cupric ammonium chloride, the two series of experiments giving closely concordant results. The heat of formation of alloys between the limits CuZn and $CuZn_2$ is considerable :—

$$CuZn \quad 46 \text{ cal. per gramme.}$$
$$Cu_2Zn_3 \; 46 \qquad ,, \qquad ,,$$
$$CuZn_2 \; 52 \cdot 5 \qquad ,, \qquad ,,$$

SPECIFIC HEAT.

The additive character of specific heat as a property is preserved in the intermetallic compounds. The early work of Regnault [119] showed that the specific heat of fusible metals followed the mixture rule at temperatures sufficiently far below their melting-point, but that as the melting-point was approached the specific heat was always greater than that calculated. The examination of a large number of alloys containing intermetallic compounds, prepared in the course of Tammann's researches, showed that the deviations from the mixture rule were less than 4 per cent.[120, 121] It was observed that the specific heat of magnesium compounds was always smaller, and that of antimony compounds greater, than that calculated by the mixture rule. The specific heat increases with the temperature, the coefficient decreasing with rise of temperature.

Some of the results are collected in the table, the deviations from the values calculated by the mixture rule having been added in the last column :—

	Sp. ht. $17\text{-}100°$.	Deviation Per Cent.		Sp. ht. $17\text{-}100°$.	Deviation Per Cent.
Al_3Mg_4	0·2301	− 1·6	Mg_3Sb_2	0·0946	− 1·6
$AgMg$	0·0884	− 3·2	Ag_3Sb	0·0553	+ 1·4
$AuMg$	0·0538	− 2·7	$CoSb$	0·0677	− 0·4
$AuMg_2$	0·0718	− 3·0	Cu_3Sb	0·0795	+ 4·2
$AuMg_3$	0·0861	− 2·3	Cu_2Sb	0·0739	+ 2·6
Cu_2Mg	0·1159	− 1·2	$ZnSb$	0·0668	+ 2·1
$CuMg_2$	0·1574	− 1·4	Ag_3Al	0·0684	0
Mg_2Si	0·2190	− 0·9	Cu_3Al	0·1068	− 1·1
$MgZn_2$	0·1156	− 1·8	$CuAl_2$	0·1496	− 0·2
$MgNi_2$	0·1255	− 5·0			

OPTICAL PROPERTIES.

A few observations of the optical properties of alloys have been made, but only a single investigation deals with the relation between these properties and the composition,[122] the latter being expressed as percentages by volume. The alloys were examined in a polished condition, and the following results were obtained :—

There is no indication whatever of a compound in the alloys of iron and nickel, or of nickel and silicon, although the occurrence of chemical combination is possible in the former, and certain in the latter instance. The indices of refraction and of absorption both bear a simple linear relation to the composition by volume. In the alloys of copper and aluminium a minimum in the absorptive index and a maximum in the refractive index correspond rather with the formula CuAl than with either Cu_3Al or $CuAl_2$. This is a series, the electrical constants of which require re-determination. In the case of the alloys of copper with nickel and with iron, which do not contain compounds, the variation of the optical constants with the composition by volume is not linear, but is similar to that of the electrical conductivity.

PHOTO-ELECTRIC PROPERTIES.

The photo-electric properties of a number of metals and a few alloys have been examined. One method is that of condensing metallic vapours on a quartz plate in a high vacuum, good mirrors being thus obtained without polishing.[123] In another investigation the metallic surfaces were smoothed by a steel scrubber working in a high vacuum.[124] The rate of photo-electric fatigue increases with the electro-positive character of the metal. The alloys of antimony and cadmium, which have been selected by many investigators on account of the very strongly marked discontinuities in the electrical properties due to the compound CdSb, have been examined in this way, and are found to exhibit a linear relation between photo-electric fatigue and composition.[125] The behaviour of alkali amalgams, and of the alloys of sodium and potassium, has also been examined with respect to their photo-electric sensitiveness.[126] The curves connecting the photo-electric sensitiveness with the wave length of the incident light are found to present maxima in the case of sodium and potassium. The liquid alloy of sodium and potassium also has a maximum, which is not a mere summation of the effects for the component metals. An examination of solid and liquid potassium amalgams showed that the selective photo-electric sensitiveness is not a function of the atom but of the molecule. On account of the complexity of the system, it is not possible to assign the effects observed to any one of the compounds of mercury with potassium. Alloys of potassium with gold, thallium, lead, antimony, and bismuth were also examined, with the result that the selective photo-electric effect was found to bear a relation to the electro-negative character of the element alloyed with the potassium. These conclusions appear likely to be of some value in a discussion of the relations between metallic alloys and electrons.

Magnetic Properties.

Whilst the magnetic properties of para- and diamagnetic alloys present few features of interest in connexion with the study of intermetallic compounds, those alloys which exhibit the property of ferromagnetism are in many respects exceptional. Of the three ferromagnetic metals, iron, nickel, and cobalt, the last two form an unbroken series of non-magnetic solid solutions at high temperatures, whilst at lower temperatures both metals undergo a change to a magnetic modification, a complete series of magnetic solid solutions existing also at low temperatures. The magnetic change is thus unaccompanied by a structural change.

The alloys of iron and nickel, on the other hand, behave in a very different manner, and the magnetic properties of these alloys have attracted the attention of many workers, the interest of the system being increased by its relation to the problem of natural meteoric irons.

As the result of several independent determinations, it has been established that the freezing-point curve of the alloys of iron and nickel has a very simple form.[127] The liquidus and solidus are separated by so small an interval as to be practically indistinguishable. There is a minimum at 67 atomic per cent of nickel, and it has been supposed that a compound, $FeNi_2$, is present, but there is no sufficient evidence of this, and the occurrence of a compound at a minimum on the freezing-point curve is otherwise without precedent. Immediately below the solidus the alloys consist of homogeneous solid solutions of γ-iron and β-nickel, which are non-magnetic. The magnetic transformation

5

temperature of iron lies at 750° and that of nickel at 320°.   At the atmospheric temperature the stable modifications are a-iron and a-nickel, both of which are ferromagnetic.   From the fact that the cooling of alloys of iron and nickel under ordinary conditions yields alloys of apparently homogeneous structure, it has been generally assumed that the two a-modifications also form a continuous series of solid solutions.   Such an assumption, however, causes great difficulties in the interpretation of the remarkable magnetic transformations undergone by these alloys.   The transformation curve falls from iron with increasing quantities of nickel, appears to reach a eutectoid point, and then rises, passing through a maximum at about 600° and 67 atomic per cent of nickel, and then again falls to the value for pure nickel.   This last part of the curve, including the maximum, does not present any special difficulties of interpretation.   The transformation temperatures obtained during heating and during cooling differ only slightly, and the curve may be fully accepted as representing an actual phase change.   The occurrence of the maximum at a composition corresponding very closely with the atomic ratio 1 Fe : 2 Ni suggests that a compound $FeNi_2$ is concerned in the change, but there is nothing to indicate whether it is present in the high temperature or the low temperature system.

The principal difficulty in the interpretation of the magnetic phenomena exhibited by these alloys presents itself in connexion with the alloys containing less than 33 atomic per cent of nickel. The temperature at which the alloys become magnetic on cooling lies much below that at which they lose their magnetic properties on heating, the difference between the two temperatures increasing with increasing percentage of nickel.   The alloys are therefore divided into " irreversible " and "reversible" alloys, the former containing less, and the latter more, than about 30 atomic per cent of nickel.[128]   The great temperature lag in the irreversible alloys renders the determination of this part of the diagram very difficult.   Its interpretation has been reached by a comparison of the artificial alloys with meteoric irons.

As already stated, artificially prepared alloys of iron and nickel consist of homogeneous solid solutions, and this is the case whether the specimens are cooled slowly or rapidly.   On the contrary, meteorites composed of iron and nickel alone contain three micrographic constituents, of which two, kamacite and taenite, are solid solutions containing about 6 per cent and 27 per cent of nickel respectively, whilst the third, plessite, is evidently a eutectic or eutectoid mixture of kamacite and taenite.[130] If we assume that the homogeneous $\beta\gamma$-solid solutions are resolved into separate constituents at a lower temperature, plessite must be regarded as a eutectoid. The eutectoid point has usually been placed at about 0°, the transformation curves obtained during cooling being taken to determine the form of the diagram. Osmond and Cartaud, however, suggested in 1904 that these curves might owe their position to undercooling, and that the curves obtained on heating might approach more closely to the position of equilibrium, thus placing the eutectoid point somewhat below 400.°

The fact that meteoric irons are converted by annealing into homogeneous solid solutions completely resembling the artificial alloys, led to the hypothesis [131] that the kamacite-plessite-taenite structure is an unstable one, not reproducible in the laboratory, and owing its origin perhaps to extremely prolonged diffusion below 0°. This somewhat improbable hypothesis has now been rendered unnecessary. The structure of meteoric iron has now been reproduced [132] by preparing an alloy with 12 per cent of nickel by the aluminothermic process, and providing for very slow cooling below 350°. Plessite is reproduced in this way without difficulty, the contrary results obtained previously being due to the annealing temperature adopted having been above 400°. The structure is far less coarse than that of the natural meteorite, owing to the slowness with which segregation can take place in a solid at so low a temperature as 350°.

It thus appears that kamacite and taenite are two limited solid solutions containing a-iron and a-nickel, and there is no conclusive evidence of the existence of a compound $FeNi_2$, although the occurrence of the maximum in the transformation curve almost exactly at that composition is at least a remarkable coincidence. The abnormalities of magnetic behaviour are no doubt accounted for by the difficulty of obtaining equilibrium in the alloys of this series. For the same reason, determinations of the electrical conductivity have hitherto given only irregular and inconclusive results.

The transformation curves in the analogous system iron-cobalt

5 *

have not been determined over the whole range of composition. The freezing-point curve is almost flat, without indication of a compound, and the magnetic transformation exhibits wide intervals between the temperatures obtained on heating and on cooling.[133]

Intermetallic compounds, one component of which is a ferromagnetic metal, are almost invariably non-magnetic, whilst solid solutions which do not contain a compound may be magnetizable in a high degree if the solvent metal be itself ferromagnetic. The following table is modified from one given by Tammann.[134] The list of compounds has been revised, only those being now included which have been determined to exist by evidence of a satisfactory kind. It should be said that the test applied is in most cases the rough one of observing whether a magnetic needle is deflected by the alloy at the ordinary temperature. The possibility is not excluded that some of these alloys may become appreciably magnetic at lower temperatures.

	Fe.	Ni.	Co.
As	Fe_2As, Fe_3As_2, $FeAs$, $FeAs_2$	Co_2As, Co_3As_2, $CoAs$	Ni_5As_2?, Ni_2As?, Ni_3As_2, $NiAs$
Bi	—	—	$NiBi$, $NiBi_3$
Cd	?	?	$NiCd_4$
Mg	—	—	Ni_2Mg, $NiMg_2$
P	Fe_3P, Fe_2P	Ni_3P, Ni_5P_2	Co_2P
Sb	Fe_3Sb_2, $FeSb_2$	$CoSb$, $CoSb_2$	Ni_3Sb, Ni_5Sb_2, $NiSb$
Sn	$FeSn_2$	Co_2Sn, $CoSn$	Ni_4Sn, Ni_3Sn, Ni_3Sn_2
Zn	$FeZn_3$, $FeZn_7$	$CoZn_4$	$NiZn_3$

Of the compounds enumerated in the above table, only Ni_2Mg and Fe_3Sb_2 exhibit appreciably magnetic properties, and it therefore appears that intermetallic compounds, one component of which is a ferromagnetic metal, are most frequently non-magnetic. This is also true of other than intermetallic compounds. Thus the majority of the oxides of the ferromagnetic metals are only very feebly magnetic, magnetic properties among the oxides of iron, for example, being almost entirely confined to the single compound Fe_3O_4, which is properly regarded as ferrous ferrite, FeO, Fe_2O_3, one of a series of magnetic ferrites in which the oxide Fe_2O_3 behaves as an acid oxide, and to one modification of ferric oxide Fe_2O_3.[135]

The borides of the metals of this group are magnetic, but not strongly so.[136]

The contrast between compounds and solid solutions in such cases as those mentioned above is very great. Solid solutions in which a ferromagnetic metal is the solvent, that is, in which it is present in by far the larger proportion, are themselves ferromagnetic, but their intermetallic compounds are only paramagnetic. This probably accounts for the difference between the amalgam of nickel[137] which is very feebly magnetic, and those of iron and cobalt[138] which are strongly magnetic.

A few series of magnetic measurements exist, from which some more quantitative conclusions may be drawn respecting these relationships. Thus, according to Honda,[139] the magnetic susceptibility of the alloys of nickel and tin lies between the susceptibilities of the two component metals, the curve being composed of a number of straight lines, each pair of which intersect at a point corresponding with the appearance of a new phase. The composition of the compound Ni_3Sn_2 is very distinctly marked in this way, but the measurements would not, in the absence of data from other sources, allow of any conclusion as to the existence of such a compound, as the limit of saturation of the solid solutions rich in nickel is equally well marked by a similar intersection. It is, in fact, the appearance of a new phase, and not the nature of that phase (compound or solid solution) which determines the appearance of a discontinuity.

The alloys of nickel and aluminium exhibit the condition somewhat more clearly. The susceptibility of the compound NiAl is about the same as that of aluminium, whilst the susceptibility of the compounds $NiAl_3$ and $NiAl_2$ is very much smaller, so that the system of intersecting straight lines passes through a minimum. This would not be likely to occur if the intermediate phases were solid solutions, not containing compounds.

Whilst the facts cited above show that the magnetic properties of a ferromagnetic element disappear wholly or partly when that element enters into combination with another metal, the more remarkable reverse case, of the formation of a ferromagnetic compound by the combination of paramagnetic or even diamagnetic elements, is also known to occur, and the discovery has opened up an entirely new field of inquiry in regard to the magnetism of alloys.

The first instance observed of the formation of a magnetic compound from non-magnetic elements is that of chromium oxide. Chromium is faintly paramagnetic; oxygen is paramagnetic or diamagnetic in its compounds, although strongly paramagnetic in the free condition. The compound Cr_4O_9 (Cr_2O_3, $2CrO_3$) was observed as far back as 1859 [140] to be strongly magnetic, whilst the same has been stated of an oxide of different composition, Cr_5O_9 ($2Cr_2O_3$, CrO_3).[141]   Other magnetic compounds of chromium probably exist.[142]

The observation was made in 1892 [143] that, whilst commercial ferro-manganese and ferro-aluminium are non-magnetic, certain ternary alloys prepared by mixing them, containing only a small proportion of iron, are very strongly magnetic. This was followed by the further discovery that the entirely non-magnetic alloys of copper and manganese are rendered magnetic by alloying with a suitable proportion of certain third metals.[144, 145] The first metal used was tin, but aluminium was soon found to produce a still greater effect. The number of investigations dealing with the Heusler alloys, as they have been called, is extremely large, and the conditions have proved to be highly complex. The principal results will be summarized here, but the complexity of the systems involved is very great, and as metastable conditions are of frequent occurrence, the properties of any Heusler alloy of given composition depend on its previous thermal and mechanical treatment. Metallographic considerations have received far too little attention in this department, and a great mass of observations has therefore accumulated, referring to alloys of unknown or doubtful internal constitution. The subject has also been encumbered with much polemical matter, dealing with questions of priority.

One fact has been established with certainty, namely, that the appearance of magnetic properties in the alloys is associated with, and dependent on, the formation of intermetallic compounds. The system which has been most fully investigated is that of copper, aluminium, and manganese. Heusler originally made the tentative suggestion [146] that a compound MnAl might be concerned, in view of the fact that the maximum of magnetic properties was found in a ternary alloy containing manganese and aluminium in approximately atomic proportions, alloyed with an excess of copper. Little stress was, however, laid on this suggestion, and the objections to it were recognized and admitted.

An explanation which at first appeared plausible was advanced by Guillaume.[147] This was to the effect that manganese was a ferromagnetic metal with a very low temperature of transformation, and that the effect of alloying with the metals in question was merely to raise the transformation point above the atmospheric temperature. This hypothesis is untenable. Manganese does not become ferromagnetic when cooled in liquid air,[148] and many other magnetic compounds have since been discovered to which such an explanation is inapplicable.

The relation of intermetallic compounds to magnetism became much more clear when the data accumulated by Heusler and his colleagues, with reference to the alloys of copper, aluminium, and manganese, were plotted on a triangular diagram, of the type generally employed in the study of ternary alloys.

The alloys of maximum magnetic properties lie on a line joining the compounds Cu_3Al and Mn_3Al. These two compounds form solid solutions with one another. Heusler and Richarz therefore proposed the following hypothesis :—[149]

"The alloys corresponding with the maximum of magnetisation . . . may perhaps be regarded as chemical compounds of the general type Al_xM_{3x}, where M represents partly Mn, partly Cu atoms in varying proportions."

Heusler and his collaborators have made constant use of this hypothesis. It is badly stated by them, in a form implying the existence of ternary compounds. For example, the maximum on the line shown in their triangular diagram occurs at the composition $2Cu_3Al$, Mn_3Al, and this is regarded by the authors as a compound $AlMnCu_2$, a member of the series $Al_x(Mn, Cu)_{3x}$. There is no evidence of the existence of such a compound. At the same time Heusler's meaning is clear from the analogy, to which he draws attention, with the double carbides of iron and manganese, $C_x(Mn, Fe)_{3x}$. These are more properly regarded as isomorphous mixtures of the carbides Fe_3C and Mn_3C, and in similar manner we may consider the magnetic alloys to be isomorphous mixtures (solid solutions) of the two definite compounds Cu_3Al and Mn_3Al. The distinction is less important than may appear at first sight. In a solid solution of progressively changing composition, molecules of the one kind are successively replaced by molecules of the other, whilst in an atomic complex of the type assumed above, the replacement is one of atoms of one kind by atoms of another. In our present state of knowledge of the internal constitution of solid solutions containing compounds, it would be difficult to distinguish effectively between the two conditions.

The form of the hypothesis suggested independently, in a slightly later paper, by Ross and Gray [150] is preferable. These authors had observed that the compound Cu_3Al, although only faintly magnetic, was more strongly so than any other alloy of the copper-aluminium series, whilst the compound Mn_3Al is known to be the most strongly magnetic member of the manganese-aluminium series.[151] It was therefore suggested that the magnetic Heusler alloys were solid solutions, containing these two compounds in varying proportions.

It is characteristic of the Heusler alloys that their magnetic properties are not fully exhibited by freshly cast specimens, but that it is necessary to subject them to an annealing process of a special kind in order that the maximum magnetic quality may appear. The most detailed experiments in this connexion are due to Take,[152] who finds that the relations between thermal treatment and magnetic properties are extremely complicated. The greatest susceptibility is obtained by quenching the alloys rapidly from a red heat and then "ageing" them at as low a temperature as possible (100°-150°). Heating to 180° for several hours, followed by slow cooling, is the most suitable thermal treatment for the alloys generally employed.[153]

The complicated nature of the magnetic transformations is due to the fact that both chemical and physical changes are involved, the nature of which is still imperfectly understood. The metallographic study of the copper-aluminium-manganese series has not yet been completed, but a detailed study of a part of the system [154] has shown the close resemblance, over a considerable range of composition, between these ternary alloys and the binary alloys of copper and aluminium. It may therefore be expected that the Heusler alloys will crystallize as homogeneous solid solutions or as two-phase systems according to the conditions of cooling. The "ageing" process may therefore be a complex one, involving (*a*) the formation, and perhaps also the decomposition, of intermetallic compounds; (*b*) the equalization of composition by diffusion ; (*c*) the union of ultramicroscopic particles to form physically more stable aggregates. It is now very commonly held that ferromagnetism is not a property of the molecule, but of molecular aggregates of notably larger size, and that the power of orientation of these magnetic complexes is very greatly modified by the character of the metal or solvent by which they are surrounded. If this be so, the complexity of the changes which occur during ageing is readily understood.

The association of magnetic properties with definite intermetallic compounds is well seen in other instances. The following compounds of manganese have been observed to be distinctly magnetic :—

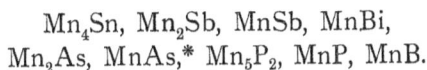

$$Mn_4Sn, \ Mn_2Sb, \ MnSb, \ MnBi,$$
$$Mn_2As, \ MnAs,^* \ Mn_5P_2, \ MnP, \ MnB.$$

Of these, the boride is strongly ferromagnetic,[155] the phosphides somewhat less so. Certain regularities have been observed. Thus, the transformation point, or temperature above which the magnetic properties disappear, rises with increasing atomic weight in the compounds of manganese with the elements of the fifth group :—[156, 157]

MnP	$18°\text{-}25°$
MnAs	$45°\text{-}50°$
MnSb	$310°\text{-}320°$
MnBi	$360°\text{-}380°$

These determinations have been made by using alloys of the composition corresponding with the formula given. Manganese nitrides are also magnetic. The bismuth compound is of special interest, as bismuth is a strongly diamagnetic element. The addition of so small a quantity as 0·25 per cent of manganese to bismuth renders it sufficiently magnetic to be attracted by a large horseshoe magnet.

Several compounds of this class are also remarkable for their possession of permanent magnetic properties. The coercive force of the boride and antimonides is not only greater than that of soft iron, but actually exceeds that of the tungsten steel used for permanent magnets. The permanent magnetism of man-

---

* A later determination by thermal methods[155] seems to show that the compounds MnAs and Mn_2As are not themselves magnetic, but that a third compound, Mn_3As_2, is formed during cooling by a reaction in the solid state, and the maximum of magnetic properties corresponds with the composition of this compound.

ganese boride is about one-half that of steel. Honda[139] has studied the dependence of permanent magnetic quality on constitution in the case of the manganese-tin alloys, and has found that the permanent magnetization attains a maximum value in the pure compound Mn_4Sn, and is rapidly lowered by small additions of either manganese or tin.

With the exception of chromium boride, compounds of this class not containing manganese have not hitherto been observed to be magnetic. Vanadium forms magnetic oxides and sulphides, but its alloys have not yet been studied from this point of view.

The paramagnetism of alloys is greatly inferior in interest to the exceptional cases of ferromagnetism, but even here the influence of intermetallic compounds is to be traced. The paramagnetism of a series of alloys is directly proportional to their composition when they consist of a conglomerate of two solid phases, but a discontinuity occurs at any composition at which a new solid phase makes its appearance, as in the cases of Ni_3Sn_2, Ni_3Sn, $NiAl_3$, and $NiAl_2$, mentioned above (p. 69).

The alloys of bismuth with tellurium and with thallium are all diamagnetic. The susceptibility curve of the former series is given as continuous,[158] but the number of alloys examined was too small to make this quite certain, and it is possible that on examination of a larger number of alloys of intermediate composition a small cusp might be found at the composition of the compound Bi_2Te_3. The bismuth-thallium susceptibility curve, however, has a distinct cusp at the composition Bi_5Tl_3.

Hall and Nernst Effects.

The Hall effect, or the transverse electromotive force which is produced by the displacement of equipotential lines in a plate of metal through which a current is flowing longitudinally, when a magnetic field is applied, has been determined for a few series of alloys. The effect has been observed to bear a relation to the thermo-electric power, and a close similarity between the relations of the two properties to the composition has been found in the two cases examined in which compounds are present. These are the alloys of antimony with cadmium, and of antimony with zinc.[83] The curves for the former series are shown in Fig. 12. The compound SbCd is most conspicuously indicated, but the second compound Sb_2Cd_3 is apparently without influence on the form of the curve, although it must be admitted that the number of alloys examined was not sufficiently large to prove conclusively that there is no small discontinuity. The antimony-zinc alloys give a curve of precisely the same form, with an enormous rise of the effect near to the compound SbZn, and no apparent indication of the second compound Sb_2Zn_3.

The value plotted as ordinates is the Hall constant, which is, however, subject to a correction which would change its absolute value, but without altering the form of the curve.

The Nernst effect, or the *E.M.F.* developed when a plate of metal through which heat is flowing is brought into a magnetic field, so that the lines of magnetic force are perpendicular to the plane of the plate, has been determined for the same alloys in the course of the same investigation. This effect also is a maximum for the compounds SbCd and SbZn in their respective series.

Comparing together the curves in Fig. 12 for the antimony-cadmium alloys, and the precisely similar group of curves obtained for the alloys of antimony and zinc, it is clear that a close relation exists between the various electrical properties of an intermetallic compound, a fact which is of high importance in connexion with the electronic theory of metallic conduction.

CRYSTALLINE FORM.

Most of the metals crystallize in the cubic system, the remainder, with the exception of tin, being hexagonal. It has been shown by Barlow and Pope [196] that the axial ratios of the hexagonal elements approximate more or less closely to those of the closest-packed assemblage of equal spheres, namely, to $a : c =$ 1 : 1˙6330 or 1 : 1˙4142. Binary compounds composed of two elements of equal valency have been shown by the same authors to crystallize also in the cubic or hexagonal system, but usually in a class which does not possess the highest symmetry of the system. Very few of the intermetallic compounds have been subjected to crystallographic measurement. This is very largely due to the difficulty of obtaining perfect crystals. When an alloy of the exact composition of an intermetallic compound is prepared, it usually forms a compact mass which breaks irregularly, without yielding definite crystals. On the other hand, crystals isolated from an alloy containing an excess of one or other component by pouring off a fusible eutectic, or by treatment with chemical reagents, are most commonly striated or deformed to such an extent that satisfactory goniometric measurements are impossible. The data as to crystalline form are consequently very scanty, and complete reliance cannot always be placed in the axial ratios. Moreover, it appears that some intermetallic compounds have a tendency to assume the appearance of a higher symmetry than they actually possess. For example, the crystals of SbSn which are a conspicuous constituent of many bearing metals have the appearance of cubes, and have been described as such, but they are really crystals of lower symmetry, as the apparent cube angles prove on measurement to differ from 90°. The following intermetallic compounds (including some native arsenides) have been examined crystallographically and described * :—

CUBIC SYSTEM.

$NaCd_2$. Combinations of octahedron and rhombic dodecahedron.
Mg_2Sn. Octahedra.
Ag_2Te. Hessite. Many observed forms.
PbTe. Altaite. Mostly massive, but with cubic cleavage.
$CoAs_2$. Smaltite ⎫
$NiAs_2$. Chloanthite ⎬ Pyritohedra.
$CoAs_3$. Skutterudite ⎪
$PtAs_2$. Sperrylite ⎭

TETRAGONAL SYSTEM.

$AuSn_2$ $a : c = 1 : 1\cdot1937$

ORTHORHOMBIC SYSTEM.

Cd_3Sb_2 $a : b : c = 0\cdot7591 : 1 : 0\cdot9687$
$FeSb_2$ $a : b : c = 0\cdot5490 : 1 : 1\cdot1224$
Ag_3Sb Dyscrasite. Pseudohexagonal
         combinations $a : b : c = 0\cdot5775 : 1 : 0\cdot6718$

* Except where other references are given, the data for artificial crystals are taken from Groth's *Chemische Krystallographie* (1906), and those for minerals from Dana's *Mineralogy* (1892).

ZnSb		$a:b:c = 0\cdot7609:1:0\cdot9598$
$FeAs_2$	Löllingite	$a:b:c = 0\cdot6689:1:1\cdot2331$
$NiAs_2$	Rammelsbergite	
$CoAs_2$	Safflorite	$\}$ Axial ratios not determined
$CeAl_4$		$a:b = 0\cdot7706:1$
$LaAl_4$		$a:b = 0\cdot7517:1$
$ThAl_4$		$a:b = 0\cdot5139:1$

HEXAGONAL SYSTEM.

CuSn	No data given	
$FeZn_7$	Resembles corundum	$a:c = 1:1\cdot2760*$
NiAs	Niccolite	$a:c = 1:0\cdot8194$
NiSb	$\{$Breithaupite	$a:c = 1:0\cdot8586$
	$\{$Artificial	$a:c = 1:1\cdot2940$
Bi_2Te_3	Tetradymite. Rhombohe-	
	dral	$a:c = 1:1\cdot5871$
Ni_2Te_3	Melonite	

MONOCLINIC SYSTEM.

$FeAl_3 \quad a:b:c = 1\cdot5413:1:1\cdot9158, \beta 107° 41'$

* From measurements by Mr. A. Scott in the author's laboratory.

CHAPTER VII.

THE EXISTENCE OF INTERMETALLIC COMPOUNDS IN THE LIQUID STATE.

THE question of the possible existence of intermetallic compounds in liquid alloys is very closely related to that of the existence of hydrates in aqueous solutions, although the former problem presents certain special experimental difficulties of its own. A consideration of the facts as a whole suggests the great probability of such a continued existence of compounds in the liquid state. In accordance with the principle of mass action, a compound which may at certain temperatures dissociate into its components must, under conditions of equilibrium, be accompanied by its products of dissociation, even if in minute quantity. The conditions of electrical conductivity in solid solutions (p. 45) point to the probability that solid intermetallic compounds are more or less dissociated, and the degree of dissociation must increase with increasing temperature. There is, however, no reason to assume that the dissociation becomes complete at the melting-point, and there is in fact much experimental evidence to show that undissociated molecules are present, although in diminished numbers, even at much higher temperatures. There is at any given temperature, an equilibrium of the form—

$$A_m B_n \rightleftarrows mA + nB$$

and this holds good of the liquid as well as of the solid state.

In the first place, certain conclusions may be drawn from the form of the freezing-point curve in binary series when the curve exhibits a maximum. Were a compound to melt entirely without dissociation, the addition of one of the components to the molten compound should (assuming that solid solutions are not formed) depress the freezing-point in accordance with Raoult's law, and the descending branch of the freezing-point curve should approximate to a straight line. As this reasoning applies to the addition of either of the constituents A and B to the compound $A_m B_n$, it follows that the ideal form of the maximum in the case of an entirely undissociated compound would be an acute intersection of two straight lines. Whether such a maximum is possible in reality has been disputed. On the one hand it has been argued [159] that a freezing-point curve must, whenever the solid and liquid phases are identical in composition, have a horizontal tangent, which excludes the possibility of a sharp angle at the maximum. On the other hand, the latter condition has been claimed to exist in binary mixtures of methyl iodide and pyridine,[160] and of iodine and chlorine at the composition ICl,[161] whilst the theoretical possibility has also been maintained, apparently on insufficient grounds.[162] It is impracticable to decide the question experimentally, as the unavoidable errors of the thermal method leave it uncertain whether the ascending and descending branches of the curve intersect or pass into one another continuously within a very short range of composition. All metallic alloys hitherto examined depart in an unmistakable manner from this ideal case, the rounding of the maximum being very perceptible. The extent of the rounding is a measure of the degree of dissociation of the compound at its melting-point.[163] It may be made quantitative by determining the normal depression of freezing-point, and comparing it with the observed curve. This is most simply done by adding to a liquid mixture having the composition of the compound an inert substance, that is, one of known molecular weight, which dissolves in it and crystallizes from it without forming either a compound or a solid solution. This method has been applied with some success to a number of systems composed of organic substances, the results indicating, for example, that the additive compound of aniline and phenol is dissociated to the extent of 20 per cent at its melting-point, and the compound of phenol and picric acid to 27 per cent.[164] The normal depression may also be calculated from the heat of fusion when this is known, using van't Hoff's formula. This method has not hitherto been applied to metallic alloys, although its application should be possible in many instances if the assumption of the validity of Raoult's law in such cases should prove to be justified.

The influence of the dissociation of the compound in the liquid phase on the solidification of the system has been exhaustively discussed from the theoretical point of view by Roozeboom.[165] The possible cases are, however, mostly such as do not present themselves in the study of alloys, and the

graphical methods used in the discussion, although of considerable interest, do not allow of the solution of any of the problems
mentioned above with the data now available.

Amongst alloys, the sharpest maxima observed are due to
compounds of magnesium, especially Mg_3Sb_2, and it must therefore be assumed that these compounds dissociate to an exceptionally small extent on melting, a conclusion which is in
accordance with their behaviour in the solid state.

The fact that these and other intermetallic compounds are
capable of existing in the liquid phase in a largely undissociated
condition at the melting-point renders it certain that undissociated molecules must also exist, although in smaller number,
at higher temperatures, the degree of dissociation increasing with
the temperature. There is also direct evidence to the same
effect. In the alloys of aluminium with antimony the compound AlSb melts above 900°, but is only formed very slowly
from its components in the liquid state, so that in one experiment only three-fourths of the quantity of compound theoretic-
ally obtainable was found in the solid alloy, after the component
metals had been heated together for thirty minutes at 1100°.[166]
It is possible that this sluggishness may have been partly due to
imperfect mixture of the two metals, the experimental method
used being open to objection, whilst aluminium frequently forms
globules, covered with a thin pellicle of oxide, which obstinately
resist union with other metals, forming an emulsion. Making
allowance for this fact, however, it appears probable that the
reaction between aluminium and antimony in the liquid state
is a slow one, molecules of the compound AlSb being formed
gradually, and persisting in the liquid condition.

The measurement of electrolytic conductivity, which has been
so constantly employed in the study of aqueous and other solutions, is not applicable to liquid alloys, in which electrolytic
conduction has never been observed to occur. Experiments to
determine the point showed a complete absence of electrolytic
conduction, even at high temperatures.[167] Moreover, there is no
evidence of any gradual transition from metallic to electrolytic
conduction.

On the other hand, molten alloys conduct metallically, and
some very remarkable results, have been obtained by determining the variation of conductivity with concentration at

a given temperature as well as the temperature coefficient of the conductivity. The only experiments of this kind hitherto published are those of K. Bornemann and his collaborators.[168, 169] The method employed was the determination of the fall of potential between two intermediate points when a current was passed through a column of the molten alloy, enclosed in a tube of glass, silica, or magnesia, according to the temperature required. Electrodes of iron or carbon were generally used, but in the case of measurements at high temperatures with such metals as nickel which attack carbon, it was found necessary to use metal electrodes, internally cooled by water. The curve exhibiting the relation between composition and electrical conductivity at a given temperature is in many cases, such as that of alloys of lead and tin, practically rectilinear, but in other cases, such as the alloys of copper and nickel, or of sodium and potassium, the curve has the deep U-form characteristic of solid solutions, alloys near the middle of the series thus having a conductivity which is very much lower than that which they would have if the property were an additive one. The reason for the occurrence of these two types of curves among homogeneous liquid alloys is not very clear, as the differences which distinguish the crystalline conglomerates of lead and tin from the solid solutions of copper and nickel no longer exist in the liquid state.

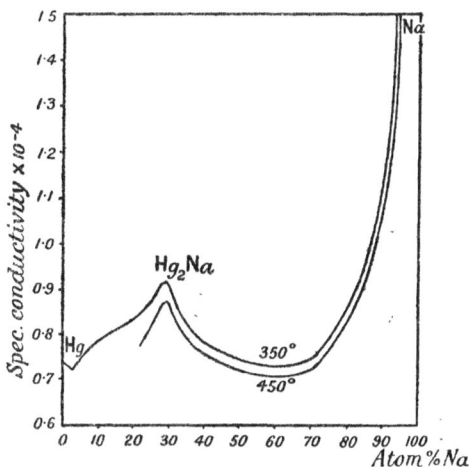

Fig. 15.

The presence of an intermetallic compound in an undissociated condition in the liquid alloy makes itself evident, as in the case of solid alloys, by a discontinuity in the curve. The best example of this is the compound Hg_2Na, which is known from other evidence to be exceptionally stable,* and is marked by

* See p. 8.

a distinct peak on the conductivity-composition curve (Fig. 15).
The two curves shown represent determinations at 350° and 450°
respectively, and the position of the peak on each curve clearly
points to the existence of the compound Hg_2Na in an undissociated condition at the two temperatures considered. The
remaining compounds of sodium and mercury which are known
to exist in the solid state are insufficiently stable at higher temperatures to produce any marked effect on the conductivity.
The corresponding compound in the potassium-mercury series,
Hg_2K, appears to be less stable than its sodium analogue, in
accordance with its lower melting-point, and it is in consequence
less distinctly marked on the conductivity curve, the peak being
considerably flattened. An unstable compound, dissociating
below its melting-point, naturally has no influence on the liquid
conductivity, and there is thus no break in the curve of the
potassium-sodium alloys at the composition Na_2K.

With liquid as with solid alloys, the manner in which the
conductivity varies with the temperature is intimately connected
with the constitution of the alloys, and the temperature coefficient curve may be employed in place of the conductivity
curve for the purpose of determining the presence or absence of
compounds. As in
the case of solid sol-
utions, the discontin-
uities due to this cause
may be even more
strongly marked in
the temperature co-
efficient than in the
conductivity itself.
Fig. 16 represents
the conditions in
liquid alloys of cop-
per and antimony.
The coefficient
plotted is that of
the specific resistance,

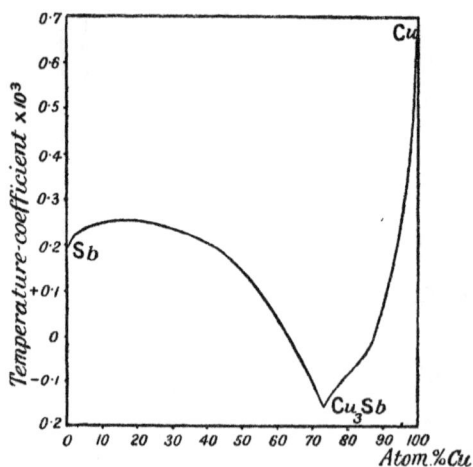

FIG. 16.

multiplied by 10^3. The very sharp change in direction of the
curve occurs at a composition in close agreement with the
formula Cu_3Sb. A minimum occurs at about the same com-

position in the conductivity curve, but there is no abrupt discontinuity, and it would not be possible from that curve alone to determine whether the compound was present in an undissociated condition, as minima have also been observed in alloys which do not contain a compound.

One remarkable feature in the curve shown in Fig. 16 deserves attention. The temperature coefficient is negative throughout a considerable range of composition in the neighbourhood of the compound Cu_3Sb. A negative coefficient of the resistance, that is, a diminution of the resistance with increasing temperature, is characteristic of electrolytic conductors, whilst the coefficient of metallic conductors is normally positive. There is, however, no reason to assume electrolytic conduction in this case. The effect is sufficiently accounted for by the progressive thermal dissociation of the solid compound Cu_3Sb with rise of temperature. Dissociation into copper and either free antimony or, more probably, the second compound Cu_2Sb, undoubtedly takes place. Alloys in the immediate neighbourhood of the compound Cu_3Sb exhibit the following peculiarity. At low temperatures, owing to dissociation of the compound, the coefficient is negative, but at higher temperatures the concentration of the dissolved undissociated compound is too small to neutralize the normal positive coefficient which characterizes alloys in general. Alloys of this series, lying within a certain range of temperature, thus possess a temperature of inversion, above which the coefficient is positive, whilst it is negative below that point. The temperature of inversion is found by experiment to lie at $1000°$ for an alloy containing 66 atomic per cent of copper (that is, 88 molec. per cent Cu_3Sb) and at $1100°$ for an alloy containing 90 atomic per cent Cu, or 40 molec. per cent Cu_3Sb. It is probable that the negative temperature coefficient of the resistance of molten cuprous sulphide is to be explained in a similar manner by simple thermal dissociation rather than by electrolytic dissociation.

Other physical properties of liquid alloys furnish similar, if less precise, indications. The diffusivity of various metals in mercury has been measured,[170] and when the atomic diffusivity is plotted as a function of the atomic weight, the values for zinc, cadmium, tin, and lead are found to lie on a smooth curve. The metals of the alkalies and alkaline earths, together with thallium, also yield points which lie on a smooth curve, but the diffusivity is throughout lower than in the former case. The characteristic difference between the two groups is that the former, from the evidence of the freezing-point curves, do not combine chemically with mercury, whilst the alkali metals and their companions form one or more compounds with mercury. The lower rate of diffusivity is thus to be accounted for by the presence of compounds in solution, so that each atom of the dissolved metal bears attached to it one or more atoms of mercury.

A very similar effect is observed in the surface tension of amalgams.[171] The surface tension of mercury is reduced by the addition of sodium, potassium, rubidium, and caesium, raised by that of lithium, calcium, and barium, and scarcely affected by that of zinc, cadmium, thallium, gold, tin, or lead. With the exception of thallium and possibly of gold, the grouping is the same as in the case of the diffusivity. The active metals are those which form compounds, whilst the inactive metals are chemically indifferent. Both series of experiments confirm the presence of undissociated molecules of intermetallic compounds in the liquid amalgams.

THE VAPOUR PHASE OF METALLIC SYSTEMS.

Determinations of the vapour density of alloys are completely lacking, and only a single instance of a definite intermetallic compound having an apparent existence in the gaseous condition is recorded. This is the compound of magnesium and zinc, $MgZn_2$, the existence of which has been proved by the thermal and microscopical methods. When mixtures of these two metals, containing an excess of zinc, are heated together in a good vacuum, crystals of the compound $MgZn_2$ are condensed on the cooler parts of the glass vessel employed.[172] The temperature of the experiment is not given, but the Jena glass tube used to contain the materials showed signs of collapse in one case, and the temperature may therefore be assumed to have been about 750°.

The fact that crystals of the compound are obtained by condensing the vapour does not prove the presence of undissociated gaseous molecules of the compound. It is possible that, although magnesium is less volatile than zinc, the two metals may distil as a mixture, the compound being re-formed by combination of the vapours when the temperature falls, the liquid alloy thus obtained subsequently crystallizing. This point might be tested by performing a series of distillations at different temperatures and pressures, and determining whether a distillate of constant composition is obtained.

Studies of the vapour-pressure relations of alloys have only been carried so far as to yield practical methods for the separation of metals or the isolation of intermetallic compounds, and the data are not available for the construction of a complete concentration - temperature - pressure diagram in even a single case. In the absence of solid solutions, the vapour pressure of a binary series at a given temperature may be expected to fall suddenly when the concentration of the more volatile component falls to that corresponding with a compound which is stable at that temperature. The case is then completely analogous with that of hydrated salts. That this condition actually presents itself is shown by the success, in a number of instances, of the method of removing an excess of a volatile metal by distillation at constant temperature, leaving a residue composed of a definite compound (see p. 30). The isolation of a compound of gold and cadmium by distilling off an excess of cadmium has been mentioned as an example of this process.

CHAPTER VIII.

THE RELATIONS OF INTERMETALLIC COMPOUNDS TO CARBIDES, SILICIDES, ETC.

CERTAIN non-metallic elements are capable of forming compounds of distinctly metallic character with metals, and these elements may therefore enter into the composition of alloys. The transition from the compounds in question to those considered in the previous sections is a perfectly gradual one, and it is impossible to make any sharp distinction between intermetallic compounds and the binary compounds of metals with elements of pronounced electronegative character. The non-metals which have the greatest tendency to form compounds having metallic properties with the metals are boron, carbon, silicon, titanium, phosphorus, arsenic, selenium, and tellurium. Of these, silicon, arsenic, and tellurium habitually form compounds of such distinctly metallic character that they may be grouped for the present purpose with the metals. Nitrogen, oxygen, sulphur, and possibly hydrogen may be regarded as alloy-forming elements under certain special conditions.

In general, the metallic properties of compounds of this kind are most pronounced when the metallic element is present in a large proportion. With phosphorus, for example, when a metal forms a series of phosphides, their metallic properties diminish as the atomic proportion of phosphorus increases. The binary mixtures, for example, of copper with the first phosphide, Cu_3P, are true alloys, but with increasing phosphorus the metallic properties of the mixtures diminish.

The alloy-forming properties of oxygen and sulphur are extremely limited. The lower oxides of copper and nickel, however, form true alloys with the respective metals, and the binary systems $Cu - Cu_2O$ and $Ni - NiO$ are strictly comparable with metallic eutectiferous systems. The eutectic structures are remarkably well developed in these two cases. As the proportion of oxide in the alloys increases, the brittleness increases and the conductivity falls, until the pure oxides are unmistakably non-metallic. Other oxides are usually either insoluble or soluble to a very limited extent in the molten metals, but other cases of miscibility may possibly be found to exist.

Copper, lead, and silver alloy with their sulphides, retaining their metallic character when the proportion of sulphide is not too large. Similarly, the selenides Cu_2Se, PbSe, and Ag_2Se form alloys with the corresponding metals. Arsenides and tellurides are included in the tabular summaries in the next section.

The nitrides are rarely metallic, only the compounds of the heavy metals of groups VI, VII, and VIII having that character. Their alloys have not been examined, but iron appears to form an alloy with nitrogen, in which the compound Fe_5N_2 is doubtless present. Titanium, chromium, and manganese also absorb nitrogen to form solid solutions. The nitrides described as possessing metallic properties are Mn_3N_2, Mn_5N_2, Mn_7N_2, CrN, Mo_3N_2, Co_2N, Fe_5N_2, Fe_2N, FeN. The nitrides of manganese are ferromagnetic. Compounds derived from ammonia by the replacement of hydrogen by the lighter metals are entirely non-metallic, whilst compounds containing the group N_3, and derived from hydrazoic acid, behave as true salts. Excluding these compounds, the only metallic nitrides which are stable at the ordinary temperature are those formed by metals of the A sub-groups of the periodic system, the metals of the B sub-groups yielding nitrides which are only stable in liquid air, and explode below the atmospheric temperature.[173] The nitrides of manganese are stable above a red heat.[174]

The electrical conductivity of the nitrides of magnesium, calcium, and aluminium is very small, whilst the conductivity of the nitrides of chromium and manganese is of the same order as that of the pure metals.[175]

The borides exhibit a considerable resemblance to the nitrides, but are much more stable. It is mainly the metals of the last three groups in the periodic system which form stable metallic borides, but even in the earlier groups some stable borides, which conduct electricity, have been prepared, including Mg_3B_2, CaB_6 and AlB_2.

The metallic borides, the formulæ of which are not yet well established, are

CrB, Mo_3B_4, WB_2, MnB, MnB_2, Fe_2B, FeB, FeB_2, Co_2B, CoB, CoB_2, Ni_2B, NiB, and NiB_2.

The carbides of the alkali and alkali earth metals and those of aluminium and the rare earth metals are readily decomposed by water, yielding various hydrocarbons. On the other hand, the carbides of some of the heavier metals are extremely stable, and form true alloys with the metals. These stable carbides include :—

VC ; Cr_4C (?), Cr_3C_2 ; Mo_2C, MoC ; W_2C, WC ; U_2C ; Mn_3C ; Fe_3C, Fe_2C ; Ni_3C.

It is possible that several other carbides of iron exist, capable of alloying with iron.

THE CHEMICAL NATURE OF INTERMETALLIC COMPOUNDS.

From the point of view of structural inorganic chemistry, the intermetallic compounds constitute at present a somewhat obscure group. Their classification and systematic treatment present special difficulties, from the want of consistency of the observed composition with the current conceptions of valency. The principle of a constant valency, fixed and invariable for each element, has long been abandoned by inorganic chemists, but such variations of valency as occur in oxides, halides, and metallic salts follow certain fairly simple rules, and conspicuous anomalies are rare. When metals combine with one another, however, much greater irregularities are observed. There is no reason to suppose that the law of definite proportions loses its validity in this class of compounds. It is generally possible to assign to an intermetallic compound a formula which only involves small numbers of atoms, the few formulæ of a high order (such as $Na_{12}Hg_{13}$) which are recorded in recent memoirs being in all probability erroneous. Nevertheless, although only relatively simple atomic ratios occur, those ratios are, in perhaps the majority of cases, irreconcilable with the valencies usually assigned to the metals. Abandoning, therefore, any attempt to determine the structure of intermetallic compounds, at least until certain preliminary problems have been solved, it will be well to consider what regularities of composition are to be observed, and the direction in which an explanation of the anomalous position of these compounds is to be sought.

The imperfection of our experimental data, to which repeated allusion has been made in these pages, renders a theoretical survey difficult. In order that a systematic comparison may be made, it is necessary that a sufficient number of binary metallic systems shall have been investigated with such completeness as to establish the number and composition of all the compounds which may be formed in each of them under the conditions of thermal analysis. This condition is far from being fulfilled. Many metallographic investigations which have been published in detail are of the nature of preliminary surveys, carried out with insufficient quantities of material and with unduly rapid rates of cooling and comparatively crude methods of observation. The diagrams of equilibrium presented as the result of such studies are therefore first approximations, probably correct in the main so far as regards the freezing-point curve, but subject to many errors in all transformations in which solid phases are involved. This circumstance, which is at once evident when a critical survey of the literature of the subject is made, has been too often overlooked, but it is of fundamental importance for any attempt at generalization. Even the careful and exhaustive revision to which the published data have been subjected by Guertler[179] and Bornemann[180] leaves our knowledge of the greater number of binary metallic systems very imperfect, and it is almost certain that many of the accepted formulæ are erroneous. Only a few systems (copper-zinc, copper-tin, copper-aluminium, silver-zinc) have been investigated with the requisite degree of thoroughness, in such a way as to embrace transformations in the solid state, and it is significant that it is precisely these systems which have been found to present the most complex conditions, a fact which suggests that the apparent simplicity of other systems may be often a mere consequence of our ignorance. In the following brief survey, an attempt has been made to utilize only those formulæ which have been established on satisfactory evidence as data for discussion. The anomalies in respect to valency, alluded to above, are undoubtedly real, and are not likely to be lessened by an experimental revision of the data, although it is highly probable that such a revision would further the discovery of a general law of composition.

As the result of an examination of the data then available, Tammann[181] enunciated two rules, to both of which exceptions may occur :—

1. Neighbouring elements in a natural group (a sub-group of the periodic system) do not form compounds with one another.
2. An element either forms compounds with all the members of a natural group, or with none of them.

The members of the first two short periods in the periodic classification were provisionally excluded from these rules, so that each natural group was regarded as consisting ordinarily of three members. In a later paper,[182] the same author supported these conclusions by a quantity of further evidence.

As regards the first rule, its scope is actually wider than is stated above. The elements of a natural sub-group, whether neighbouring or not, do not form compounds with one another, the only recorded exception being the compound of iodine with bromine. The rule at once breaks down if it is extended to include the elements of the first two short periods. For instance, magnesium forms compounds with zinc, cadmium, and mercury, whilst sodium combines with potassium and probably also with the succeeding elements of its group.

The second rule is subject to numerous exceptions. The extent of its applicability is discussed below.

According to Mendeleeff's conception of the periodic system, the sum of the valencies of any element towards oxygen and towards hydrogen is equal to 8. It was suggested by Kurnakoff[183] that the types of combination observed in the alkali amalgams might be accounted for in this way. The valency of the alkali metals towards oxygen is only 1, and a maximum valency of 7 towards elements differing widely from oxygen, such as other metals, might therefore be expected. The results of a thermal investigation of amalgams were regarded as confirming this view, but the rule has not been found of use in practice.

The relations between electro-affinity and valency have been considered by Abegg.[184] The effective valency of an element depends on the nature of the element (or group) with which it is in combination, and is the more variable the greater the difference between the elements concerned. This difference is expressed by the electro-affinities, or, in the periodic classification, by the horizontal distance between the two elements in the table. Compounds built up of two widely separated elements are heteropolar, and those of closely neighbouring elements homopolar. The one class passes continuously into the other, but the distinction is clear when it is seen that potassium chloride is a typically heteropolar compound, whilst iodine chloride is an extreme instance of homopolar combination. The intermetallic compounds are obviously of the latter type, and the application of this conception to the data collected by Tammann has been made by Abegg.[185] This author distinguishes between normal and contra-valencies, the maximum number of which is shown in the table below :—

Group.	I	II	III	IV	V	VI	VII	VIII
Normal valencies	+ 1	+ 2	+ 3	$\Big\}\pm 4\Big\{$	− 3	− 2	− 1	$\Big\}\pm 8$
Contra-valencies	(− 7)	(− 6)	(− 5)		+ 5	+ 6	+ 7	

Negative electro-affinity is in general much weaker than positive, and this is particularly evident in the contra-valencies, hence the numbers in brackets are maximum values, not often actually attained.

Two elements of the same natural group will only combine together if the change of electro-affinity with atomic weight is considerable. This is mainly the case with the elements of low atomic weight (members of the two short series). Hence aluminium combines with boron and silicon with carbon, but not with the higher members of the respective sub-groups.

The more strongly heteropolar any pair of elements may be, the more probable is it that they will combine with one another in accordance with their maximum normal valencies, whilst homopolar pairs are more likely to combine in varying proportions, differing widely from the normal valencies. The intermetallic compounds are typical representatives of the latter class.

The transition from one class to the other may be illustrated by the example of the phosphides. Phosphorus forms compounds of distinctly heteropolar character with strongly positive metals, such as potassium and calcium, these phosphides having the properties of weak salts, hydrolysed by water to a metallic hydroxide and hydrogen phosphide. In contrast with these, the weakly positive metals such as iron, nickel, and copper, form phosphides of distinctly metallic character, which may be fairly classed among alloys.

Tammann's first rule finds its justification in the absence of any marked difference of electro-affinity between members of the same sub-group. The change of electro-affinity in the two short series is, however, considerable, and it is therefore natural to expect that the elements belonging to these series will combine with the elements of higher atomic weight belonging to the same group. This conclusion may now be examined in the light of the more complete data which have accumulated since the publication of the papers to which reference has been made.

It is not necessary to assume that the combining power of the metals with one another finds its full expression in the results of thermal analysis. A compound is only recognized by this method when it is capable of existing as a distinct phase, in equilibrium with other phases. Abegg has pointed out, however, that such an independent existence presupposes that the compound in question reaches so high a concentration in the solution (liquid or solid) that saturation occurs. When the affinity between two homopolar elements is small, combination may indeed take place, but the concentration of the compound thus formed may not reach the limit of saturation, so that a distinct phase does not make its appearance. Improvements in the methods of determining the electrical conductivity and thermo-electric properties, by throwing light on the constitution of homogeneous phases, may make the detection of combination possible even in these cases.

Group I.—Systems of two alkali metals have been little examined. Lithium, in its alloy-forming properties, resembles magnesium rather than sodium, and its behaviour is in some respects anomalous. Molten lithium is hardly miscible with sodium or potassium.[186] Sodium forms a single compound, Na_2K, with potassium, but this compound dissociates below its melting-point.[187] It is probable that both lithium and sodium would prove, on investigation, to form compounds with rubidium and caesium, but that the metals of the potassium sub-group would not combine with one another.

In the second sub-group, copper, silver, and gold alloy together, forming solid solutions either completely or to a limited extent. Sodium forms a single compound with gold, $NaAu_2$, melting without decomposition at 989°,[188] whilst silver crystallizes in a pure state from its solution in molten silver.[188]

Group II.—The alloys of beryllium have not been studied. Magnesium alloys readily with most metals, and is remarkable for its power of forming compounds which melt with extremely little dissociation, but which are frequently highly reactive towards air or moisture. Hardly anything is known of alloys containing strontium, barium, or radium. Magnesium forms a single stable compound with calcium, Ca_3Mg_4.[189]

The second sub-group is composed of typical alloy-forming metals. Zinc does not form compounds with either cadmium or mercury, and cadmium also alloys with mercury without combination. Magnesium, however, combines with all three members of the sub-group, forming the compounds:—

MgCd (see p. 16).

$MgZn_2$, with the high melting-point of 595° and a compound with mercury, of unknown composition.

Group III.—Boron, although a non-metal, forms compounds of a distinctly metallic character with some of the heavy metals. These have already been considered (p. 87).

Scandium, yttrium, and the remaining tervalent metals of the rare earths have not been investigated in regard to their alloy-forming properties, and the same may be said of gallium. Indium and thallium form an incomplete series of solid solutions. Aluminium is not miscible with thallium in the liquid state.

Group IV.—The carbides and silicides have been considered in a previous section. Carbon silicide (carborundum) CSi, is typical of the transition from intermetallic compounds to compounds of the class of sulphides and oxides.

Nothing is known of the alloys of titanium, zirconium, or germanium, or of the alloys of cerium with any of the metals of this group.

Group V.—The mutual behaviour of vanadium, niobium, and tantalum is unknown. The elements nitrogen, phosphorus, arsenic, antimony, and bismuth form a highly natural group, in which the electro-affinity changes throughout, at first rapidly and then more slowly.

Arsenic and antimony form solid solutions within the limits yet investigated (o to 50 atomic per cent As) and the series will probably prove to be complete.[190] Antimony and bismuth form a complete or almost complete series of solid solutions.[191]   It does not appear that arsenic combines with bismuth on fusion, and there are even indications that only limited miscibility occurs in the liquid state.[192]

Group VI.—Alloys of chromium, molybdenum, tungsten, and uranium with one another have not been investigated.   The group sulphur, selenium, tellurium is a natural one, with a progressive increase of metallic properties with increasing atomic weight.   The oxides, sulphides, selenides, and tellurides of the elements of the sixth group are entirely non-metallic in character, and are not to be classed amongst alloys.

Group VII.—Manganese is the only known metallic element in this group.   An exception to Tammann's first rule occurs amongst the non-metallic members of the group, iodine forming a compound with its neighbour bromine.   In accordance with the rapid change of electro-affinity in the two short series, fluorine and chlorine combine readily with the succeeding members.

Group VIII.—This group includes three sub-groups, each composed of three closely allied elements.   The members of each sub-group form a continuous series of solid solutions with one another.   In the iron group the relations are somewhat complex, as there is some evidence of chemical combination occurring between iron and nickel, and between iron and cobalt, although the conditions of existence of the compounds have not been definitely determined.   The isomorphism of the two sub-groups of platinum metals respectively appears to be perfect.

As regards the relations of the three sub-groups to one another, iron and platinum form alloys which undergo somewhat complex changes in the solid state.[193]   Cobalt and nickel form solid solutions with platinum and palladium, but the limits of concentration are quite unknown.

Compounds of Metals of Group I with the Metals of Other Groups.

The alkali metals of Group IA, have been mainly investigated, for obvious reasons, in their relations to metals of comparatively low melting-point, and their combinations with the less fusible metals are little known.   Taking the second group first, neither sodium nor potassium forms a compound with magnesium, whilst each combines with zinc to form a compound, provisionally represented as $NaZn_{11}$ or KZn_{11}, although the actual number of zinc atoms is uncertain. The compounds with cadmium include $LiCd$, $LiCd_2$, $NaCd_2$, $NaCd_6$, and potassium compounds of doubtful composition. The alkali amalgams are remarkable for the number and variety of the compounds they contain, and for the high melting-point of the principal compound in each series :—

Li_3Hg	Na_3Hg	KHg	$RbHg_6$	Cs_2Hg ?
Li_2Hg ?	Na_5Hg_2	KHg_2		$CsHg$?
$LiHg$	Na_3Hg_2	KHg_3		$CsHg_2$
$LiHg_2$	$NaHg$	K_2Hg_9 ?		$CsHg_4$
$LiHg_3$	Na_7Hg_8 ?	KHg_{10} ?		$CsHg_6$
	$NaHg_2$			$CsHg_{10}$?
	$NaHg_4$			

the compound RHg_2 being the most stable one in each case.

Sodium and potassium do not combine with aluminium, and the only compounds of alkali metals with metals of Group III are those with thallium, Na_5Tl_2 ?, Na_2Tl ?, $NaTl$; K_2Tl ?, and KTl.

Tin and lead combine with the alkali metals, and the following compounds are described :—

Li_4Sn	Na_4Sn	Na_4Pb	K_2Sn ?	K_2Pb ?
Li_3Sn_2	Na_2Sn	Na_3Pb	KSn ?	KPb_2 ?
Li_2Sn_5	Na_4Sn_3	$NaPb$	KSn_2	KPb_4
	$NaSn$	Na_2Pb_3	KSn_4	
	$NaSn_2$			

In the fifth group, the nitrides, phosphides, and arsenides of the alkali metals are in no respect metallic. They contain almost invariably three atoms of the alkali metal united with one atom of the element of the higher valency, although combination may also take place in other proportions, as in the case of the compounds derived from hydrazoic acid, HN_3, which, however, have all the properties of true salts.

An approach to the character of intermetallic compounds is seen in the antimonides and bismuthides :—

Li_3Sb	Na_3Sb	K_3Sb
	$NaSb$	

	Na_3Bi	K_3Bi
	$NaBi$	K_3Bi_2
		KBi ?
		KBi_2

In the sixth group, the sulphides, selenides, and even tellurides, are not to be regarded as forming alloys, although the equilibrium diagrams closely resemble those of binary metallic mixtures.   In most instances several compounds are formed.

The sub-Group IB, consisting of the metals copper, silver, and gold, differs so widely from the alkali sub-group that close analogies are not to be expected.   The following compounds with metals of the second group are well established :—

Cu_2Mg	$AgMg$	$AuMg$
$CuMg_2$	$AgMg_3$	$AuMg_2$
		Au_2Mg_5
		$AuMg_3$
Cu_4Ca	Ag_4Ca	
	Ag_2Ca	
	Ag_3Ca_2	
	$AgCa$	
$CuZn$	$AgZn$	$AuZn$
Cu_2Zn_3	Ag_2Zn_3	Au_2Zn_5
$CuZn_2$?	Ag_2Zn_5	$AuZn_8$?
$CuZn_6$?		
Cu_2Cd	$AgCd$	$AuCd$
Cu_2Cd_3	Ag_2Cd_3	$AuCd_3$
	$AgCd_3$?	

The nature of the compounds with mercury is still unknown.

Of compounds with metals of Group III, only those with aluminium are known :—

Cu_4Al ?		Au_4Al ?
Cu_3Al		Au_5Al_2
$CuAl$	Ag_3Al	Au_2Al
$CuAl_2$	Ag_2Al	$AuAl$
		$AuAl_2$

Thallium does not combine with copper, silver, or gold.

A few regularities in the above series may be observed. Compounds in which one atom of the univalent metal is united with two atoms of the bivalent metal are of frequent occurrence, and generally exhibit the greatest stability of the series :—

$LiCd_2$, $LiHg_2$; $NaCd_2$, $NaHg_2$; KHg_2; $CsHg_2$; $CuMg_2$, $CuZn_2$?; $AuMg_2$.   A silver compound of this type has not been recorded, and there is some doubt whether $CuZn_2$ or Cu_2Zn_3 more correctly represents the γ-phase of the copper-zinc alloys, there being on this point a conflict between the thermal and the electrical evidence (see pp. 20 and 52).

The compounds with elements of Group IV are of a very diversified character. Copper forms a series of silicides of a metallic nature, but only Cu_3Si has been definitely established. Silver does not combine with silicon, but the two elements crystallize separately from their molten alloys. With tin the compounds are :—

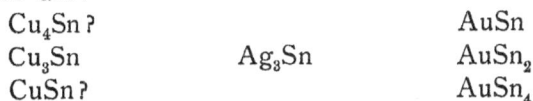

Cu_4Sn ?		$AuSn$
Cu_3Sn	Ag_3Sn	$AuSn_2$
$CuSn$?		$AuSn_4$

Copper and silver do not combine with lead, whilst gold forms the compounds Au_2Pb and $AuPb_2$.

With Group V the difference of electro-affinity becomes so great that there is a conspicuous tendency to form compounds in correspondence with the normal valencies. The phosphides and arsenides have not been completely studied, owing to the volatility of one component, but the series as far as known may be represented thus :—

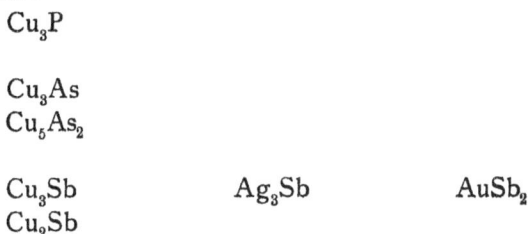

Cu_3P

Cu_3As
Cu_5As_2

Cu_3Sb	Ag_3Sb	$AuSb_2$
Cu_2Sb		

Bismuth does not combine with any of the metals of this subgroup.

In Group VI, the sulphides Cu_2S and Ag_2S possess semi-metallic properties, which become more strongly marked in the selenides Cu_2Se and Ag_2Se. The tellurides are more complex, and form true alloys. The following are known : Cu_2Te, Cu_4Te_3 ; Ag_2Te, $AgTe$; $AuTe_2$. In accordance with its high atomic weight and tendency to assume a higher valency, the compounds of gold are much less regular than those of copper and silver.

The only member of Group VIB, the behaviour of which in this respect has been examined, is chromium, which does not combine with copper or silver, whilst manganese perhaps froms a

7

compound Ag_2Mn.   As regards Group VIII, there is a well-marked tendency to form solid solutions with the copper group, and there is no proof of the existence of any compound.

Compounds of the Metals of Group II with Metals of Other Groups.

Magnesium is remarkable for its power of forming intermetallic compounds of great thermal stability, which dissociate very little on fusion.   Calcium has been little investigated, and, on account of experimental difficulties, little reliance is to be placed in the published formulæ of most of its compounds with metals.   With Group III are formed :—

$$Mg_4Al_3$$

Mg_8Tl_3?	
Mg_2Tl	$CaTl_3$
Mg_3Tl_2	Ca_3Tl_4
	$CaTl$

A definite metallic silicide, Mg_2Si, exists, and the following compounds with tin and lead have been described :—

Mg_2Sn	$CaSn_3$?
	Ca_2Pb
	$CaPb$
Mg_2Pb	$CaPb_3$

In Group V, the compounds Mg_3Sb_2 and Mg_3Bi_2 are well established, and Mg_3As_2 is highly probable.   Nothing is known of alloys of magnesium with chromium or manganese.   In the eighth group, magnesium combines with nickel, forming the compounds Mg_2Ni and $MgNi_2$, and probably with the platinum metals, but quantitative data are lacking.

The alloys of the metals of the sub-Group IIB have been the object of much investigation, but the data are even now very incomplete, especially in regard to the amalgams.

Zinc combines with aluminium, forming a compound which probably has the composition Zn_3Al_2, but cadmium and aluminium do not alloy, and the nature of aluminium amalgam is unknown. Thallium does not combine with zinc or cadmium, but forms a compound Hg_2Tl with mercury.

No compounds of metals of Group IIB with those of Group IVB exist.   The compounds with Group V are as follows :—

Zn_3P_2	Cd_3P_2?	Hg_3P_2?
Zn_3As_2	Cd_3As_2	
	$CdAs_2$	
Zn_3Sb_2	Cd_3Sb_2	
ZnSb	CdSb	

Bismuth does not combine with any of the metals of the sub-group.

The sulphides and selenides of the zinc metals have the simple formulæ ZnS, etc., but are hardly at all metallic. The tellurides ZnTe, CdTe, and HgTe are well defined, and combination does not take place in other proportions. Chromium and manganese undoubtedly combine with the metals of the zinc group, but the formulæ of their compounds are unknown.

In the eighth group complex series of compounds occur, of which the following have been established with more or less certainty :—

Zn_7Fe
Zn_3Fe

Zn_3Ni Cd_4Ni
ZnNi

Zn_4Co?

The platinum metals form numerous compounds with zinc, cadmium, and mercury, but their formulæ have not been definitely established in a single case.

COMPOUNDS OF THE METALS OF GROUP III WITH METALS OF OTHER GROUPS.

Aluminium does not combine with silicon, tin, or lead, and indium forms a complete series of solid solutions with lead. Aluminium forms the compounds Al_4Ce, Al_2Ce, AlCe, $AlCe_2$, and $AlCe_3$ with cerium.[104] Thallium does not combine with silicon or tin, and the doubtful possibility of a compound with lead has been previously discussed (p. 11).

Aluminium forms the compound AlSb, but does not combine with bismuth. Thallium appears to form the compound Tl_3Sb, in which its resemblance to the alkali metals is apparent, and also the compounds Tl_3Bi and Tl_3Bi_5? Further, the com-

7 *

pounds $AlCr_3$, Al_3Mn? and $AlMn_3$ have been recorded.   Aluminium certainly combines with the less fusible metals of Group VI, but the statements as to the nature of the compounds formed are uncertain and conflicting.

The following compounds have been recorded in Group VIII :—

Al_3Fe?	Al_3Ni	Al_4Co?
	Al_2Ni	Al_5Co_2?
	$AlNi$	$AlCo$

Thallium does not combine with any metals of the iron group. Of compounds of aluminium and thallium with the platinum metals, the only representatives definitely known are Al_3Pt and TlPt.

COMPOUNDS OF THE METALS OF GROUP IV WITH METALS OF OTHER GROUPS.

For the present purpose, silicon must be included amongst the metals so far as its compounds with the less fusible metals are concerned.   The following compounds of tin and lead with the elements of Groups V, VI, and VII may be tabulated :—

Sn_3As_2	Pb_3As_4?
SnAs	
$SnAs_2$?	
Sn_3Sb_2	
SnSb	

Neither tin nor lead combines with bismuth.

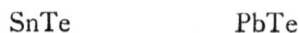

SnSe	PbSe
Sn_2Se?	
SnTe	PbTe

Chromium does not combine with tin or lead.   Manganese does not unite with lead, but forms a series of compounds with silicon and tin :—

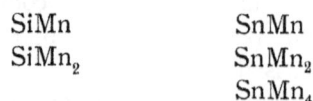

SiMn	SnMn
$SiMn_2$	$SnMn_2$
	$SnMn_4$

The conditions in Group VIII are fairly complex :—

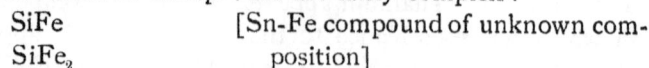

SiFe	[Sn-Fe compound of unknown com-
$SiFe_2$	position]

Si_3Ni_2
SiNi
Si_2Ni_3?
$SiNi_2$
$SiNi_3$?

Sn_2Ni_3
$SnNi_2$
$SnNi_4$

Si_3Co
Si_2Co
SiCo
Si_2Co_3?
$SiCo_2$

SnCo
$SnCo_2$

Sn_3Pt
Sn_3Pt_2
SnPt
$SnPt_3$

Pb_2Pt
PbPt

Pb_2Pd
PbPd
$PbPd_2$
$PbPd_3$

Lead does not combine with the metals of the iron group, and the silicides of the platinum metals are not very well known.

COMPOUNDS OF THE ELEMENTS OF GROUP V WITH THE METALS OF SUCCEEDING GROUPS.

The phosphides are included in the tabular summary below :—

		Sb_2Se_3	Bi_2Se_3
	As_2Te_3	Sb_2Te_3	Bi_2Te_3
PCr?		Sb_2Cr	[Bi and Cr do
		SbCr	not combine]
PMn?	AsMn	Sb_3Mn_2?	
P_3Mn_5	$AsMn_2$	$SbMn_2$	
PFe_2	AsFe	Sb_2Fe	[Bi and Fe do
PFe_3	As_2Fe_3	Sb_2Fe_3	not combine]
	$AsFe_2$		
PNi_2	AsNi	Sb_3Ni_2	
P_2Ni_5	As_2Ni_3	SbNi	BiNi
PNi_3	As_2Ni_5	Sb_2Ni_5	$BiNi_3$
		$SbNi_3$	

PCo_2	$AsCo$	Sb_2Co	
	As_2Co_3	$SbCo$	
	$AsCo_2$		
	As_2Co_5		
	$As_2Pt?$	Sb_2Pt	$Bi_2Pt?$
	As_3Pt_2	$SbPt$	
		Sb_2Pt_5	
		Sb_2Pd	Bi_2Pd
		$SbPd$	
		Sb_3Pd_5	
		$SbPd_3$	

COMPOUNDS OF THE METALS OF GROUP VI WITH THOSE OF GROUPS VII AND VIII.

Very few of the tellurides have been studied, the only formula established with any certainty being $PtTe_2$. Tellurium combines with iron, nickel, and cobalt, and probably also with the other platinum metals, but the composition of the compounds is unknown.

Chromium enters into solid solution with manganese, iron, cobalt, and nickel, and the formation of compounds has been suspected, but not yet proved. Its alloys with the platinum metals have received little attention.

The relations of molybdenum and tungsten to the metals of the eighth group appear to be very complex, but little reliance can be placed in the formulæ hitherto published for such compounds, based as they are mainly on the chemical analysis of residues. Thus, an extraordinarily complex series of compounds of molybdenum with manganese and iron have been described,[105] but with little justification.

Manganese is the solitary metallic member of Group VII, and, whilst it readily forms solid solutions with the members of Group VIII, there are no definite indications of the formation of compounds.

The data summarized in this section may now be briefly reviewed from the theoretical standpoint, as far as their imperfect character permits.

In the first place, Tammann's second rule, that an element of one sub-group forms compounds with all the elements of another sub-group or with none of them, may be tested in its application to intermetallic compounds. One conspicuous exception was noticed by Tammann himself. Lead does not combine with copper, and does not even form a homogeneous liquid with it at the melting-point. It alloys readily with silver, but again without chemical combination. On the other hand, lead and gold combine together, forming two definite intermetallic compounds.

Further examples of the same kind may be noted. Whilst nearly all the metals form antimonides, a much smaller number combines chemically with bismuth. Aluminium combines with zinc, but not with cadmium, whilst thallium does not combine with either zinc or cadmium, but does so with mercury. Manganese and also copper combine with tin, but apparently not with lead. The exceptions all tend in the same direction, the readiness to form compounds increasing with the heteropolarity of the two elements concerned.

Abegg remarks, however, that the data of the thermal method, being obtained at relatively high temperatures, fix only a lower limit to the combining power of the metals, as the stability of labile homopolar compounds, to which class most intermetallic compounds belong, may be expected to diminish with increasing temperature. The more exact thermal methods which have been devised since the publication of Abegg's paper have made possible also the detection of compounds which are formed in the solid state at comparatively low temperatures, but the best results are to be expected from physical determinations, performed on alloys which have been brought to a condition of equilibrium by very prolonged annealing below the temperature at which combination is suspected to take place. Such compounds with a low temperature range of stability may prove very important in filling the gaps and explaining the anomalies of the periodic arrangement.

In view of the uncertainty that prevails as to the composition of the majority of intermetallic compounds, it would be futile to attempt any structural representation of their constitution. Reference must, however, be made to one or two suggestions which have been put forward in this connexion. The intermetallic compounds found in certain amalgams, and containing a number of mercury atoms associated with a single atom of a metal usually regarded as of low valency, were considered by Kerp[51] to contain mercury of crystallization, and to be comparable with hydrated salts. Thus, $NaHg_5$ was regarded as being either $NaHg$, $4Hg$, or Na_2Hg, $9Hg$. There is here an obvious attempt to obtain formulæ consistent with the ordinary valencies, sodium being univalent, whilst mercury is represented as univalent in the first and bivalent in the second formula. This view is so far justified that there is an undoubted resemblance in the equilibrium diagrams of many alloys and hydrated salts. A liquidus curve with several maxima is obtained in such systems as ferric chloride-water, whilst the decomposition of an inter-metallic compound on heating into a new solid phase and a mother-liquor, such as :—

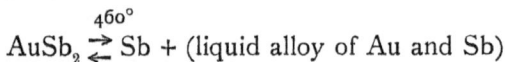

$$AuSb_2 \overset{460°}{\rightleftharpoons} Sb + \text{(liquid alloy of Au and Sb)}$$

is completely comparable with the decomposition of many hydrates on heating, for example :—

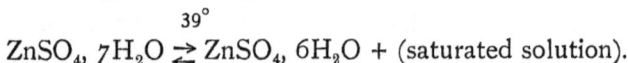

$$ZnSO_4, 7H_2O \overset{39°}{\rightleftharpoons} ZnSO_4, 6H_2O + \text{(saturated solution)}.$$

The constitution of hydrated salts is, however, still so obscure that the analogy is not directly very helpful, although it points to the probable advantage of considering the co-ordination numbers of the metals, to which reference is made below.

The next view is that which regards metals as uniting with one another by means of their contravalencies or latent valencies. This is the view adopted by Abegg. It does not appear to be applicable in a quantitative sense. Thus, whilst silver and mag-nesium, according to their positions in the periodic system, have the normal valencies + 1 and + 2, and the contravalencies – 7 and – 6 respectively, this fact does not enable us to understand why the only two compounds which are formed under the con-ditions of thermal analysis should have the formulæ $AgMg$ and $AgMg_3$. It also fails to account for such compounds as $CaZn_{10}$ and $CsHg_{10}$, in which the number of atoms of the one metal associated with a single atom of the other exceeds the maximum number of contravalencies.

It is probable that light may be thrown on the constitution of the intermetallic compounds by a careful study of their crystal-line form. The important hypothesis of Barlow and Pope[196] establishes a relation between the crystallographic constants of a compound and the valencies involved in its formation from its component elements. The hypothesis in its original form has proved of value chiefly in its application to organic compounds, but it has been successfully applied to some binary inorganic compounds.[197] Recently, however, it has received an important modification or extension by the work of Barker,[198] whose attempt to explain some unusual types of isomerism may have far-reaching theoretical consequences. It has been shown that certain pairs or groups of isomorphous compounds, the chemical formulæ of which as arranged in accordance with the ordinary valencies of the elements exhibit no similarity, may be reduced to a similar type by formulation in accordance with Werner's co-ordination theory. For example, the isomorphous pairs :—

(1) $K_2SnCl_4, 2H_2O$ and K_2FeCl_5, H_2O

(2) $MnCl_2, 4H_2O$ and $BeNa_2F_4$

do not present any obvious similarity of composition in their ordinary formulation, but when formulated as co-ordinated compounds their analogies are at once apparent :—

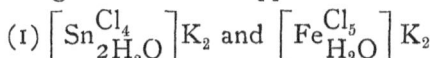

$$(1) \left[Sn^{Cl_4}_{2H_2O} \right] K_2 \text{ and } \left[Fe^{Cl_5}_{H_2O} \right] K_2$$

(2) $[Mn, 4H_2O]Cl_2$ and $[BeF_4] Na_2$

The test of isomorphism cannot be applied to intermetallic compounds until a much larger body of evidence as to their crystallographic characters has been accumulated. The two very closely analogous compounds Mg_2Sn and Mg_2Pb are not completely isomorphous, but form two series of solid solutions separated by a gap.[199] Complete miscibility has been observed between the following pairs : Cu_3Al and $CuZn$; Cu_2Zn_3 and the γ-constituent of the copper-aluminium series : Cu_2Zn_3 and $NiZn_3$.

Evidently the available data are far too scanty to admit of any immediate conclusions, but the method indicated is one which is likely to be fruitful.

No systematic attempt to devise a nomenclature for the intermetallic compounds has been made. Investigators have in general been content to denote the compounds by their formulæ. Cooke, in 1855, used the terms "stibiotrizincyl" and "stibiodizincyl" for the supposed compounds $SbZn_3$ and $SbZn_2$ respectively,[13] but his example has not been followed. Kurnakoff has frequently employed the terms "mercuride," "plumbide," "stannide," etc., based on an analogy with "silicide," "arsenide," and "antimonide". Thus $CsHg_6$ is caesium hexamercuride, Mg_2Sn is magnesium stannide, etc. This example, which suggests the possibility of a systematic nomenclature, has not been generally followed.* The introduction of mineralogical names for the separate micrographic constituents of non-ferrous alloys derived from the "austenite," "sorbite," etc., of steels, is to be deprecated, as leading to an accumulation of trivial names, and standing in the way of a scientific nomenclature.

*This plan has now been adopted in the large "Lexicon der anorganischen Verbindungen" in course of publication under the editorship of M. K. Hoffmann (Leipzig).

CHAPTER IX.

TERNARY COMPOUNDS.

THE number of ternary intermetallic compounds hitherto recorded is very small, although it is possible that others may be discovered when more ternary systems have been investigated with sufficient attention. In general, however, it appears to be possible to predict the character of a ternary system from a consideration of its component binary systems, and new phases, peculiar to the ternary system, do not in such cases make their appearance. Ternary compounds should be looked for in systems containing strongly heteropolar elements.

Two such compounds have been recorded from amalgams.[200] A maximum is found on the freezing-point surface of the cadmium-sodium amalgams at 325°, corresponding with a compound CdHgNa. This may perhaps be regarded as derived from $NaHg_2$ by the replacement of one atom of mercury by the closely allied cadmium, or as a double compound, $NaCd_2$, $NaHg_2$. Similarly, in the sodium-potassium amalgams a maximum has been observed at 188°, corresponding with a compound Hg_2KNa. This is most conveniently regarded as a double compound of two members of the corresponding binary series, NaHg and KHg.

Both of these compounds have been described on the evidence of cooling curves taken in the immediate neighbourhood of the maxima in question, and in each case the initial freezing-point was found to be lowered by the addition of the components. The complete liquidus surfaces have not yet been described.

The third compound of this class occurs in the system aluminium-magnesium-zinc, and has a more complex composition.[201] A plane section through the space-model, having as base a line joining the two compounds Al_3Mg_4 and $MgZn_2$, is shown in Fig. 17. The thermal analysis of alloys falling within this range of composition indicates the formation of a compound, melting with decomposition at 505°, and forming solid solutions

with the aluminium-magnesium compound.   The maximum heat
of reaction

compound $\rightleftarrows$ liquid and solid solution

at 505° occurs at the composition—

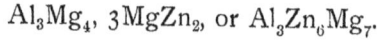

Al_3Mg_4, $3MgZn_2$, or $Al_3Zn_9Mg_7$.

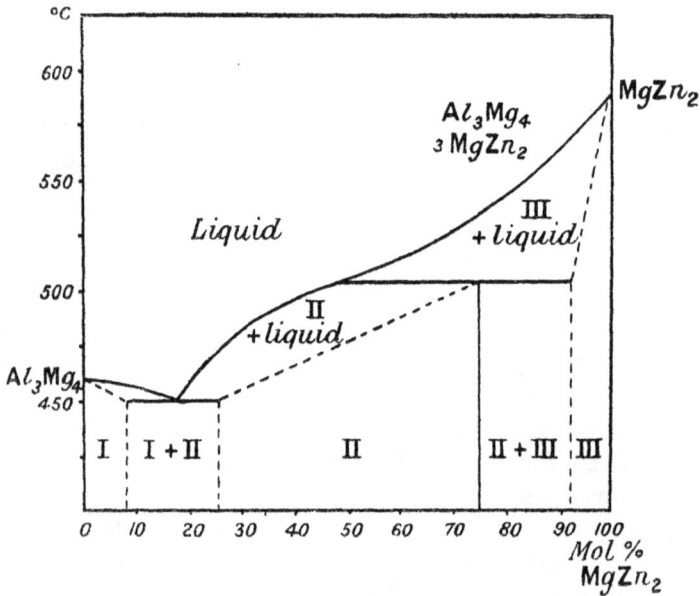

Fig. 17.

The maximum brittleness is also found to coincide with this
composition.   The micrographic examination fails in this part
of the system, on account of the brittleness of the alloys and of
the very slight differences in electrolytic character between constituents present in the same section.

A closer study of similar systems will probably reveal the
existence of other ternary compounds, but any discussion of
their constitution would be premature.

REFERENCES.

[1] F. Rudberg, *Pogg. Ann.*, 1830, **18**, 240; 1831, **21**, 317.

[2] A. Levol, *Ann. Chim. Phys.*, 1852, [iii], **36**, 193; **39**, 163.

[3] F. Guthrie, *Phil. Mag.*, 1875, [iv], **49**, 1.

[4] F. Rüdorff, *Pogg. Ann.*, 1864, [v], **2**, 337.

[5] F. Guthrie, *Phil. Mag.*, 1884, [v], **17**, 462.

[6] A. Gorboff, *J. Russ. Phys. Chem. Soc.*, 1909, **41**, 1241.

[7] Karsten, *Pogg. Ann.*, 1839, [ii], **16**, 160.

[8] Crookewit, *Annalen*, 1848, **68**, 289.

[9] F. Crace Calvert and R. Johnson, *Phil. Mag.*, 1855, [iv], **10**, 240.

[10] *Phil. Trans.*, 1858, **148**, 349, and later papers.

[11] A. Matthiessen, *ibid.*, 1858, **148**, 369, and later papers.

[12] *Brit. Assoc. Report*, 1863, 37.

[13] J. P. Cooke, *Amer. J. Sci.*, 1855, [ii], **20**, 222.

[14] N. S. Kurnakoff, S. F. Schemtschuschny, and V. Tararin, *Zeitsch. anorg. Chem.*, 1913, **83**, 200.

[15] C. T. Heycock and F. H. Neville, *Phil. Trans.*, 1897, **189** A, 25.

[16] *Ibid.*, 1900, **194** A, 201.

[17] *Ibid.*, 1902, **202** A, 1.

[18] F. H. Neville, *Brit. Assoc. Report*, 1900.

[19] C. H. Desch, *J. Inst. Metals*, 1909, **1**, 227.

[20] See, for example, H. Gautier, *L'Etude des Alliages* (Paris, 1901).

[21] H. W. B. Roozeboom, *Zeitsch. physikal. Chem.*, 1899, **30**, 385.

[22] Ponsot, *Bull. Soc. chim.*, 1895, [iii], **13**, 312. Compare the earlier work of Rüdorff and de Coppet.

[23] G. G. Urazoff, *Zeitsch. anorg. Chem.*, 1909, **64**, 375.

[24] R. Kremann, *Monatsh.*, 1904, **25**, 1271.

[25] M. Chikashigé, *Zeitsch. anorg. Chem.*, 1906, **51**, 328.

[26] D. P. Smith, *ibid.*, 1907, **56**, 109.

[27] G. Tammann, *ibid.*, 1903, **37**, 303; 1905, **45**, 24; 1905, **47**, 289.

[28] F. M. Jaeger, *Zeitsch. Kryst.*, 1904, **38**, 555.

[29] J. H. Adriani, *Zeitsch. physikal. Chem.*, 1900, **33**, 453.

[30] K. Lewkonja, *Zeitsch. anorg. Chem.*, 1907, **52**, 452.

[31] N. S. Kurnakoff and N. A. Pushin, *ibid.*, 1907, **52**, 430.

[32] N. S. Kurnakoff and S. F. Schemtschuschny, *ibid.*, 1909, **64**, 149.

[33] S. F. Schemtschuschny, *ibid.*, 1906, **49**, 400.

[34] E. S. Shepherd, *J. Physical Chem.*, 1904, **8**, 421.

[35] W. Reinders, *Zeitsch. anorg. Chem.*, 1900, **25**, 113; F. E. Gallagher, *J. Physical Chem.*, 1906, **10**, 93.

[36] N. S. Konstantinoff and W. A. Smirnoff, *J. Russ. Phys. Chem. Soc.*, 1911, **43**, 1211.

[37] W. Gontermann, *Zeitsch. anorg. Chem.*, 1907, **55**, 419.

[38] *Metallographie*, i., 795.

[39] P. N. Degens, *Zeitsch. anorg. Chem.*, 1909, **63**, 207.

[40] W. Rosenhain and A. P. Tucker, *Phil. Trans.*, 1908, **209** A, 89.

[41] D. Mazzotto, *Internat. Zeitsch. Metallographie*, 1911, **1**, 289.

[42] G. Voss, *Zeitsch. anorg. Chem.*, 1908, **57**, 34.

[43] R. Vogel, *ibid.*, 1909, **63**, 169.

[44] G. G. Urazoff and R. Vogel, *ibid.*, 1910, **67**, 442.

[45] H. C. H. Carpenter, *Internat. Zeitsch. Metallographie*, 1912, **3**, 170.

[46] N. S. Kurnakoff and N. S. Konstantinoff, *Zeitsch. anorg. Chem.*, 1908, **58**, 1.

[47] S. F. Schemtschuschny, *ibid.*, 1906, **49**, 384.

[48] S. F. Schemtschuschny, *Internat. Zeitsch. Metallographie*, 1913, **4**, 228.

[49] A. Schleicher, *ibid.*, 1913, **3**, 102.

[50] W. Kerp, *Zeitsch. anorg. Chem.*, 1898, **17**, 284.

[51] W. Kerp, W. Böttger, H. Winter, and H. Iggena, *ibid.*, 1900, **25**, 1.

[52] A. Guntz and Férée, *Compt. rend.*, 1900, **131**, 182.

[53] A. C. Vournasos, *Ber.*, 1911, **44**, 3266.

[54] P. Lebeau, *Compt. rend.*, 1900, **130**, 502.

[55] *Ibid.*, 1902, **134**, 284.

[56] C. van Eyk, *Proc. K. Akad. Wetensch. Amsterdam*, 1902, **10**, 859.

[57] A. van Bijlert, *Zeitsch. physikal. Chem.*, 1891, **8**, 343.

[58] W. D. Bancroft, *J. Physical Chem.*, 1902, **6**, 178.

[59] W. J. van Heteren, *Zeitsch. anorg. Chem.*, 1904, **42**, 129.

[60] O. Brunck, *Ber.*, 1901, **34**, 2733.

[61] H. C. H. Carpenter and C. A. Edwards, *Proc. Inst. Mech. Eng.*, 1907, 57.

[62] C. T. Heycock and F. H. Neville, *Trans. Chem. Soc.*, 1892, **61**, 914.

[63] C. R. Groves and T. Turner, *ibid.*, 1912, **101**, 585.

[64] W. R. E. Hodgkinson, R. Waring, and A. P. H. Desborough, *Chem. News*, 1899, **80**, 185.

[65] P. Lebeau, *Compt. rend.*, 1906, **142**, 154.

[66] M. Philips, *Metallurgie*, 1907, **4**, 587, 613.

[67] E. Rudolfi, *Zeitsch. anorg. Chem.*, 1907, **53**, 216.

[68] W. Guertler, *Metallurgie*, 1908, **5**, 184, 621.

[69] W. Guertler and G. Tammann, *Zeitsch. anorg. Chem.*, 1905, **47**, 163.

[70] *Ibid.*, 1906, **49**, 93.

[71] E. Maey, *Zeitsch. physikal. Chem.*, 1899, **29**, 119; 1901, **38**, 289, 292; 1904, **50**, 200.

[72] A. Matthiessen, *Phil. Trans.*, 1860, **150**, 177.

[73] E. van Aubel, *Compt. rend.*, 1901, **132**, 1266.

[74] S. F. Schemtschuschny, *Internat. Zeitsch. Metallographie*, 1913, **4**.

[75] R. Frilley, *Rev. de Métallurgie*, 1911, **8**, 457.

[76] N. S. Kurnakoff and S. F. Schemtschuschny, *Zeitsch. anorg. Chem.*, 1908, **60**, 1.

[77] G. G. Urazoff, *ibid.*, 1911, **73**, 31.

[78] W. J. Smirnoff and N. S. Kurnakoff, *ibid.*, 1911, **72**, 31.

[79] T. Turner and M. T. Murray, *J. Inst. Metals*, 1909, **2**, 98.

[80] A. Martens, *Mitt. K. techn. Versuchs-Amt.*, 1890, **8**, 236.

[81] W. Guertler, *Zeitsch. anorg. Chem.*, 1906, **51**, 397.

[82] N. J. Stepanoff, *ibid.*, 1912, **78**, 1.

[83] A. W. Smith, *Phys. Rev.*, 1911, **32**, 178.

[84] N. S. Konstantinoff and W. A. Smirnoff, *Internat. Zeitsch. Metallographie*, 1912, **2**, 153.

[85] N. A. Pushin and A. V. Baskoff, *J. Russ. Phys. Chem. Soc.* 1913, **45**, 746.

[86] N. A. Pushin and W. Rjaschsky, *Zeitsch. anorg. Chem.*, 1913, **82**, 50.

[87] N. A. Pushin and E. G. Dischler, *ibid.*, 1913, **80**, 65.

[88] W. Broniewski, *Ann. Chim. Phys.*, 1912, [viii], **25**, 5.

[89] J. Dewar and J. A. Fleming, *Phil. Mag.*, 1892, [v], **34**, 326; 1893, [v], **36**, 271.

[90] W. Nernst, F. Koref, and W. A. Lindemann, *Sitz. k. Akad. Wiss. Berlin*, 1910, 165, 247.

[91] A. Eucken and G. Gehlhoff, *Ber. deut. physikal. Ges.*, 1912, **14**, 169.

[92] G. Wiedemann and R. Franz, *Pogg. Ann.*, 1853, [ii], **89**, 497.

[93] W. Jaeger and H. Diesselhorst, *Abh. phys.-techn. Reichsanst.*, 1900, **3**, 269.

[94] W. Haken, *Ann. Physik.*, 1910, [iv], **32**, 291.

[95] E. Rudolfi, *Zeitsch. anorg. Chem.*, 1910, **67**, 65.

[96] K. Mönkemeyer, *ibid.*, 1905, **46**, 415.

[97] H. Pélabon, *Compt. rend.*, 1908, **146**, 1397.

[98] H. Fay and H. E. Ashley, *Amer. Chem. J.*, 1902, **27**, 95.

[99] G. I. Petrenko, *Zeitsch. anorg. Chem.*, 1906, **50**, 133.

[100] R. Vogel, *ibid.*, 145.

[101] W. Biltz and W. Mecklenburg, *ibid.*, 1909, **64**, 226.

[102] M. Kobayashi, *ibid.*, 1910, **69**, 1.

[103] K. Hüttner and G. Tammann, *ibid.*, 1905, **44**, 131.

[104] H. Fay and C. B. Gillson, *Amer. Chem. J.*, 1902, **27**, 81.

[105] W. Reinders, *Zeitsch. physikal. Chem.*, 1903, **42**, 225.

[106] H. C. Bijl, *ibid.*, 1902, **42**, 641.

[107] A. P. Laurie, *Trans. Chem. Soc.*, 1888, **53**, 104.

[108] A. P. Laurie, *Phil. Mag.*, 1892, [v], **33**, 94.

[109] M. Herschkowitsch, *Zeitsch. physikal. Chem.*, 1898, **27**, 123.

[110] N. A. Pushin, *Zeitsch. anorg. Chem.*, 1907, **56**, 1.

[111] A. Sucheni, *Zeitsch. Elektrochem.*, 1906, **12**, 726.

[112] Person, *Ann. Chim. Phys.*, 1848, [iii], **24**, 129.

[113] M. P. E. Berthelot, *ibid.*, 1901, [vii], **22**, 317.

[114] *Ibid.*, 1879, [v], **18**, 433.

[115] A. Galt, *Proc. Roy. Soc. Edin.*, 1898, **22**, 137.

[116] J. H. Gladstone, *Phil. Mag.*, 1900, [v], **50**, 231.

[117] W. F. Luginin and A. Schukareff, *Arch. Sci. phys. nat.*, 1902, [iv], **13**, 5; 1903, [iv], **15**, 49.

[118] T. J. Baker, *Phil. Trans.*, 1901, **196** *A*, 529.

[119] H. V. Regnault, *Ann. Chim. Phys.*, 1841, [iii], **1**, 129.

[120] H. Schimpff, *Zeitsch. physikal. Chem.*, 1910, **71**, 257.

[121] A. V. Saposhnikoff, *J. Russ. Phys. Chem. Soc.*, 1909, **41**, 1708.

[122] J. T. Littleton, *Phys. Rev.*, 1911, **33**, 453.

[123] R. Pohl and P. Pringsheim, *Ber. deut. physikal. Ges.*, 1912, **14**, 506, 546.

[124] K. Herrmann, *ibid.*, 557.

[125] *Ibid.*, 573.

[126] R. Pohl and P. Pringsheim, *ibid.*, 1910, **12**, 215, 349, 697, 1039.

[127] R. Ruer and E. Schüz, *Metallurgie*, 1910, **7**, 415.

[128] J. Hopkinson, *Proc. Roy. Soc.*, 1890, **47**, 23.

[129] Reichenbach, *Pogg. Ann.*, 1861, **114**, 99, 250.

[130] F. Osmond and G. Cartaud, *Rev. de Métallurgie*, 1904, **1**, 69.

[131] W. Fraenkel and G. Tammann, *Zeitsch. anorg. Chem.*, 1908, **60**, 416.

[132] C. Benedicks, *Rev. de Métallurgie*, 1911, **8**, 85.

[133] W. Guertler and G. Tammann, *Zeitsch. anorg. Chem.*, 1905, **45**, 205.

[134] G. Tammann, *Zeitsch. physikal. Chem.*, 1908, **65**, 73.

[135] S. Hilpert, *Ber.*, 1909, **42**, 2248.

[136] H. Moissan, *Compt. rend.*, 1895, **120**, 173; 1896, **122**, 424.

[137] H. Wünsche, *Ann. Physik.*, 1902, [iv.], **7**, 116.

[138] H. Nagaoka, *ibid.*, 1896, [iii.], **59**, 66.

[139] K. Honda, *Ann. Physik.*, 1910, [iv], **32**, 1003.

[140] F. Wöhler, *Annalen*, 1859, **111**, 117.

[141] I. I. Shukoff, *Compt. rend.*, 1908, **146**, 1396.

[142] P. Pascal, *ibid.*, 1909, **148**, 1463.

[143] T. W. Hogg, *Chem. News*, 1892, **66**, 140.

[144] F. Heusler, *Verh. deut. physikal. Ges.*, 1903, **5**, 219.

[145] W. Stark and E. Haupt, *ibid.*, 222.

[146] F. Heusler, *Zeitsch. angew. Chem.*, 1904, **17**, 260.

[147] C. E. Guillaume, *Actes. Soc. helv. Sci. nat.* 1907, i. 88.

[148] W. Preusser, *Dissertation*, Marburg, 1908.

[149] F. Heusler and F. Richarz, *Zeitsch. anorg. Chem.*, 1909, **61**, 265.

[150] A. D. Ross and R. C. Gray, *Proc. Roy. Soc. Edin.*, 1909, **29**, 274.

[151] G. Hindrichs, *Zeitsch. anorg. Chem.*, 1908, **59**, 414.

[152] E. Take, *Abh. k. Ges. Wiss. Göttingen*, 1911, [ii], **8**, No. 2.

[153] A. D. Ross, *Trans. Faraday Soc.*, 1912, **8**, 185.

[154] W. Rosenhain and F. C. A. H. Lantsberry, *Proc. Inst. Mech. Eng.*, 1910, 119.

[155] E. Wedekind, *Ber.*, 1907, **40**, 1259.

[156] S. Hilpert and T. Dieckmann, *ibid.*, 1911, **44**, 2831.

[157] S. Hilpert, T. Dieckmann, and E. Colver-Glauert, *Trans. Faraday Soc.*, 1912, **8**, 207.

[158] C. E. Mendenhall and W. F. Lent, *Phys. Rev.*, 1911, **32**, 406.

[159] R. Ruer, *Zeitsch. physikal. Chem.*, 1907, **59**, 1; 1908, **64**, 357.

[160] A. H. W. Aten, *Proc. K. Akad. Wetensch. Amsterdam*, 1904, **7**, 468.

[161] W. Stortenbecker, *Zeitsch. physikal. Chem.*, 1889, **3**, 11.

[162] R. Kremann, *Ahrens' Sammlung*, 1909, **14**, Heft 6, 23.

[163] W. Stortenbecker, *Zeitsch. physikal. Chem.*, 1892, **10**, 183.

[164] R. Kremann, *Monatsh.*, 1904, **25**, 1215.

[165] W. H. B. Roozeboom, *Zeitsch. physikal. Chem.*, 1905, **53**, 448. See also A. Findlay and E. M. Hickmans, *Trans. Chem. Soc.*, 1907, **91**, 905.

[166] G. Tammann, *Zeitsch. anorg. Chem.*, 1905, **48**, 53.

[167] J. Kinsky, *Zeitsch. Elektrochem.*, 1908, **4**, 406.

[168] K. Bornemann and P. Müller, *Metallurgie*, 1910, **7**, 396.

[169] K. Bornemann and G. von Rauschenplat, *ibid.*, 1912, **9**, 473.

[170] G. McPhail Smith, *Zeitsch. anorg. Chem.*, 1908, **58**, 381.

[171] F. Schmidt, *Ann. Physik.*, 1912, [iv], **39**, 1108.

[172] A. J. Berry, *Proc. Roy. Soc.*, 1911, **86** A, 67.

[173] F. Fischer and F. Schröter, *Ber.*, 1910, **43**, 1465.

[174] E. Wedekind and T. Veit, *ibid.*, 1908, **41**, 3769.

[175] I. I. Shukoff, *J. Russ. Phys. Chem. Soc.*, 1910, **42**, 40.

[176] A. Binet du Jassoneix, *Ann. Chim. Phys.*, 1909 [viii.], **17**, 145.

[177] A. Moissan, *Compt. rend.*, 1894, **119**, 185.

[178] *Ibid.*, 1896, **122**, 274, 1463.

[179] W. Guertler, *Metallographie* (Berlin, in course of publication).

[180] K. Bornemann, *Die binären Metallegierungen* (Halle, in course of publication)

[181] G. Tammann, *Zeitsch. anorg. Chem.*, 1906, **49**, 113.

[182] *Ibid.*, 1907, **55**, 289.

[183] N. S. Kurnakoff, *ibid.*, 1900, **23**, 439.

[184] R. Abegg, *ibid.*, 1904, **39**, 330.

[185] *Ibid.*, 1906, **50**, 309.

[186] G. Masing and G. Tammann, *ibid.*, 1910, **67**, 183.

[187] G. L. C. M. van Rossen Hoogendyk van Bleiswyk, *ibid.*, 1912, **74**, 152.

[188] C. H. Mathewson, *Internat. Zeitsch. Metallographie*, 1911, **I**, 51.

[189] N. Baar, *Zeitsch. anorg. Chem.*, 1911, **70**, 352.

[190] N. Parravano and P. de Cesaris, *Intern. Zeitsch. Metallographie*, 1912, **2**, 70.

[191] N. Parravano and E. Viviani, *Atti. R. Accad. Lincei*, 1910, [v], **19**, i, 835.

[192] K. Friedrich and A. Leroux, *Metallurgie*, 1908, **5**, 158.

[193] E. Isaac and G. Tammann, *Zeitsch. anorg. Chem.*, 1907, **55**, 63.

[194] R. Vogel, *ibid.*, 1912, **75**, 41.

[195] G. Arrivaut, *Compt. rend.*, 1906, **143**, 285, 464.
[196] W. Barlow and W. J. Pope, *Trans. Chem. Soc.*, 1906, **89**, 1675.
[197] *Ibid.*, 1907, **91**, 1150.
[198] T. V. Barker, *ibid.*, 1912, **101**, 2484.
[199] A. von Vegesack, *Zeitsch. anorg. Chem.*, 1907, **54**, 367.
[200] E. Jänecke, *Zeitsch. physikal. Chem.*, 1906, **57**, 507.
[201] G. Eger, *Internat. Zeitsch. Metallographie*, 1913, **4**, 29.

INDEX.

8 *

ABERDEEN : THE UNIVERSITY PRESS

www.ingramcontent.com/pod-product-compliance
Lightning Source LLC
Chambersburg PA
CBHW031812190326
41518CB00006B/306